Praise for *The Secret Scie*...

"What has physics got to do with conception, birth, and infancy? Plenty indeed, as Michael Banks explains—and if you decide to find out, I guarantee you won't be disappointed. *The Secret Science of Baby* is fun and fascinating, but it is more than that: it deepens our sense of awe and admiration at the ingenuity life has put into the process of propagating itself."

—Philip Ball, author of *How to Grow a Human: Adventures in Who We Are and How We Are Made*

"If you're a parent who's ever pondered conception, childbirth, or baby behavior, and wondered, 'How the heck does this even happen?', *The Secret Science of Baby* is for you. Prepare to be wowed—and maybe a little grossed out, too."

—Melinda Wenner Moyer, author of *How to Raise Kids Who Aren't Assholes: Science-Based Strategies for Better Parenting—from Tots to Teens*

"From conception to birth and beyond, *The Secret Science of Baby* walks the reader through the wide and fascinating range of science involved in the production of a new human, and in the process makes this most quintessentially miraculous of life events seem even more amazing."

—Chad Orzel, author of *How to Teach Quantum Physics to Your Dog*

"Fascinating. Terrifying. I'll never look at human reproduction the same way again."

—James Breakwell, comedy writer, creator of @XplodingUnicorn on Twitter, and author of *Bare Minimum Parenting*

"Everything you wanted to know about the science of your bouncing, burping bundle of joy but were just too plain exhausted to ask. Michael Banks explores the fascinating and, frankly, downright weird physics, chemistry, and biology of making a human. *The Secret Science of Baby* covers everything from the pregnancy-detecting capabilities of frogs—I kid you not—to the fluid dynamics of diapers, seamlessly blending wry historical vignette with state-of-the-art research and captivating explanations of the scientific principles at play. Just like your teething little one at 3:00 am, *The Secret Science of Baby* is unputdownable."

—Philip Moriarty, author of *When the Uncertainty Principle Goes to Eleven: Or How to Explain Physics with Heavy Metal*

"*The Secret Science of Baby* offers a fascinating look at some of the scientific triumphs (and mishaps) of creating small humans. This book is perfect for anyone interested in learning about the science of early parenthood or fun trivia facts like what was used as a common feeding bottle during the Middle Ages."

—Julie Vick, author of *Babies Don't Make Small Talk (So Why Should I?): The Introvert's Guide to Surviving Parenthood*

"This was a hilarious, informative, and entirely impractical guide to babies. I loved it."

—Chris Ferrie, author of *Quantum Physics for Babies*

THE SECRET SCIENCE OF BABY

THE SECRET SCIENCE OF BABY

The Surprising Physics of Creating a Human, from Conception to Birth—and Beyond

MICHAEL BANKS

BenBella Books, Inc.
Dallas, TX

BenBella Books, Inc.
10440 N. Central Expressway
Suite 800
Dallas, TX 75231
benbellabooks.com
Send feedback to feedback@benbellabooks.com

BenBella is a federally registered trademark.

Printed in the United States of America
10 9 8 7 6 5 4 3 2 1

Library of Congress Control Number: 2022014668
ISBN 9781637741467 (trade paperback)
ISBN 9781637741474 (electronic)

Editing by Jodi Frank and Alexa Stevenson
Copyediting by Michael Fedison
Proofreading by Leah Baxter and Jenny Rosen
Indexing by Debra Bowman
Text design and composition by PerfecType, Nashville, TN
Illustrations by Michael Banks
Cover design by Faceout Studio, Amanda Hudson
Cover image © Shutterstock / Bert Flint (bear), M_Guneshi (space mobile), and Nadiinko (atom)
Printed by Lake Book Manufacturing

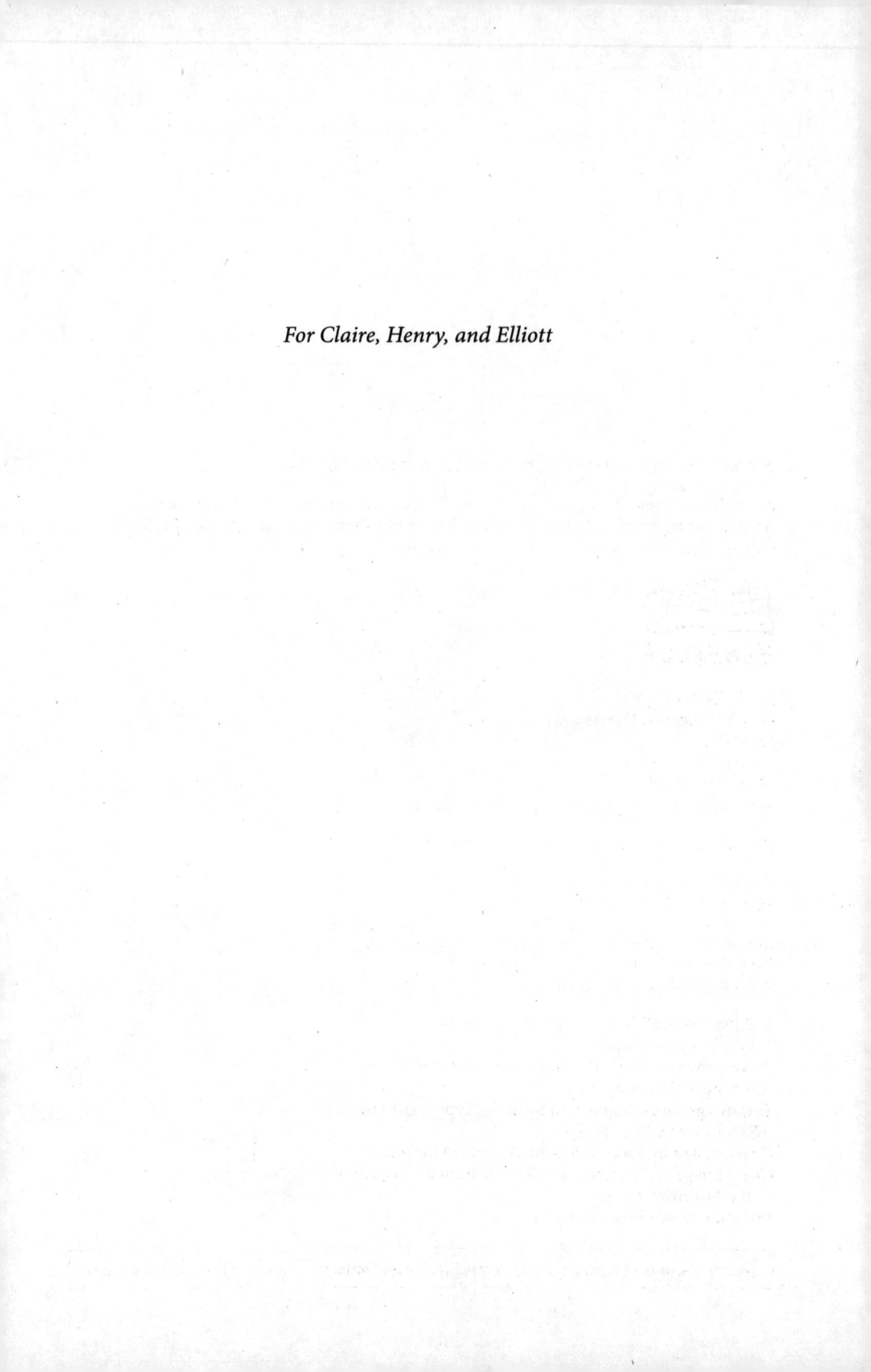

For Claire, Henry, and Elliott

CONTENTS

INTRODUCTION

If you have picked up this book, there is a good chance you either have children or have one on the way. Or, perhaps you may just be curious about how physics can possibly have anything to do with babies. If you are expecting, then congratulations! Your life will now, or soon, change forever—interrupted nights; never going out in the evening; going to the restroom with an audience; pushing your screaming child in the buggy in the rain desperate for him or her to sleep. Again, many congratulations! After blood, sweat, and plenty of tears, childbirth is a moment of pure joy and something you will most probably only experience once or twice in your life. It is a major milestone, especially for grandparents who have been waiting years, if not decades, for the moment they can finally put that homemade knitted hat on their grandchild for the first time.

Pregnancy is also a time when parents* are bombarded with information: what to expect, what not to do, and how that little embryo is developing week-to-week (including the bizarre use of different fruits as a unit of length). If you are after any of that, then, alas, you may be better served elsewhere. This book will not tell you what to eat during pregnancy or describe the techniques to get your baby to sleep through the night. Nor will it advise you

* Rather than use "parents and carers" throughout the book, I only use the term "parents" for ease of readability. There may also be instances where I use "mother" or "father" but of course recognize that there are same-sex couples who may use different terms.

how to raise a perfect violin-playing, Mandarin-speaking toddler. Research into the latest child psychology is absent here, having been well trodden in countless other books.

Instead, *The Secret Science of Baby* will explain, from a physics perspective, *how* or *why* things happen: For example, how can cells in the uterus contract in unison to expel a 4-kilogram (8-plus-pound) baby? What is it about an infant's cry that solicits parents' attention so effectively? How does a newborn effortlessly extract milk from the breast? The book will also shine a light on the latest science behind many aspects of conception, pregnancy, and baby development while not shying away from highlighting the limitations of such work and what remains to be discovered. The book will detail how such research has generated answers, new insights, and, yes, more questions.

The Secret Science of Baby also highlights how technological development over many decades has reflected the true spirit of curiosity-driven physics research. We'll explore techniques developed by physicists, engineers, and mathematicians to get a better understanding of what the human body, particularly in baby form, can do.

When people think about physics, the first thing that likely comes to mind is the universe, quantum mechanics, or the hunt for fundamental particles of nature. Yet physics is a subject that touches on so many more different areas, and many of the tools that physicists have built—be it machines like magnetic resonance imaging (MRI) or theoretical models—have often found uses beyond those directly intended. Albert Einstein's theory of general relativity in the early twentieth century, for example, paved the way decades later for global positioning systems that allow us to know where we are anywhere on the globe with a precision of a few meters (or a few yards), while the advent of quantum mechanics in the 1920s promises unbreakable communication as well as computational ability far more powerful than their "classical" counterparts. And, yes, even when it comes to conception, pregnancy, and babies, there are many insights that physics can provide, from examining how infants draw their first breath and understanding how gases diffuse through the placenta to touching upon the science of how infants acquire language.

While focusing on the physics, there may be times when I stray into other areas such as chemistry, mathematics, engineering, and even neuroscience. Indeed, the power of physics, as we will see, lies in its interdisciplinary nature. Yet, I hope to show that examining *how* things work—a perspective I did not come across when I was head deep in parenting and pregnancy books as an expectant parent—might lead to a better appreciation of how it is possible to create another human and why and how our bodies, and babies, do what they do.

After all, the infant universe is a special place.

THE COITUS EXPERIMENT

It is fall 1991, and the Dutch physiologist Pek van Andel is attending a scientific meeting on medical technologies that is being held at the University of Groningen in the Netherlands. He enters the university and sits down in one of the nondescript lecture rooms looking forward to the many presentations that will be given that day. Talk after talk passes by, but then one address piques van Andel's interest. During it, the speaker plays a video showing the inner workings of the mouth and throat of a professional opera singer belting out "aaaaaa." The black-and-white moving images of the vocal tract are incredibly clear and mesmerizing in detail, showing anatomically how the larynx of a talented singer produces such a rich sound. Working at the Groningen University Hospital at the time, van Andel is awestruck by the beautiful "body art" images he is witnessing that day.

During the rest of the conference, with the film still playing in van Andel's mind, he begins to think, as any good scientist would, how he could employ the technique for his own research.

He has an idea. Yet, this would not be his typical area of scientific expertise, and it would also require help from others. He first tries to persuade his wife, who refuses, but instead offers to speak to a friend, the organizational anthropologist Ida Sabelis. Intrigued and open-minded enough about the proposal, Sabelis, with her then-boyfriend (identified only as Jupp), agrees to meet van Andel and colleagues at the hospital to discuss the experiment. In the spring of 1992, the pair travel the three hours by train from Amsterdam to Groningen. Following a briefing about the procedure, all parties are satisfied under a few caveats, such as scans are to be done in secret and the movies are only to be used for scientific purposes. The date of the scan is then set: October 24, 1992.

As the months pass by and the day of the scan gets ever closer, Sabelis becomes more and more apprehensive while still wanting to go ahead. "I was now worried that it was going to happen and what the consequences would be," noted Sabelis later.[1] "What would my colleagues, friends, and family say?" Nevertheless, on that autumn day, the pair arrive at the hospital. They are greeted by the researchers and quickly ushered into a room. Given the tight space inside the machine, van Andel's idea would be a challenge, but Sabelis and her partner are undeterred. They listen as the researchers go through the steps of the procedure, finish their coffee, take a quick stop at the restroom, and return before getting undressed. Then they lie beside each other on a small rectangular table—barely wide enough to fit them both—before being gently slid into the heart of the scanner. After some fumbling, a change of position, and some sniggering, the couple finally *do* it.

"The erection is fully visible, including the root," a voice booms over the intercom. "Now lie down, be very still, and hold your breath during the scan." A few whirs of the machine take place and then it is back to "work." At the end of the session, the pair enter history as the first couple to have sex in an MRI machine.

Well, officially anyway.

⚛ ⚛ ⚛

The images that van Andel witnessed during the conference on that day in 1991 were not like a typical movie produced with a video camera, but one made by a revolutionary new technology called magnetic resonance imaging, or MRI. The technique was developed in the 1970s and 1980s by several people, including the UK physicist Sir Peter Mansfield, who shared the 2003 Nobel Prize in Physiology or Medicine for his contributions together with the US chemist Paul Lauterbur. MRI machines are used to safely peer through the skin noninvasively to see what lies beneath. They were, and still are, huge pieces of equipment, mostly due to the need for a rather large magnet.* During an MRI scan, the patient lies still inside the cylindrical bore of the magnet, the hollow space being only some 40 centimeters (15.8 inches) in diameter—not pleasant for anyone with claustrophobia. Coils then scan the parts of the body that need medical attention.

Some of the magnets in the first MRI machines were made of large solenoid-shaped copper coils so that when a current was driven through, it could produce a static magnetic field. Generating higher fields required more current, which became more expensive. Some modern MRI machines use superconducting magnets, which feature materials like niobium-titanium that, when cooled to a certain low temperature, conduct electricity without resistance. This means they are more efficient and require less power, allowing them to generate much higher magnetic fields. The disadvantage is that they require liquid helium to cool them to temperatures of around -240° Celsius (-400° Fahrenheit).

The role of the powerful magnet in an MRI machine is to pull on protons that are inside of us. Our bodies are about 60 percent water, which is comprised of two hydrogen atoms and one oxygen atom. The hydrogen atom's nucleus is a single positively charged proton, and when the magnetic field generated by the large magnet is switched on, this pulls the proton's "spin" to align along the same direction as the applied magnetic field—with a larger magnetic field resulting in better proton spin alignment. A way to visualize this is to imagine the earth spinning on its axis but with an arrow pointing out of it to represent the spin direction. When a magnetic field is applied along the north, the proton's spin will align along that direction.

---

* The magnetic field employed by MRI scanners is typically around 1.5 tesla, or about thirty thousand times stronger than that produced by the earth.

After the protons have been pulled into one direction, the MRI machine then directs radio waves at them. The radio waves are tuned to a specific frequency so that the protons absorb a certain amount of energy. This changes the direction of the protons' spin away from the direction of the magnetic field. When the radio waves are turned off, the protons "relax," and the speed at which they do so depends on the chemical nature of the molecules in different parts of the body. This allows contrast to occur in different tissues in the body for clearer images. Picking up these relaxation signals via gradient coils located around the MRI machine enables 3D imaging to occur. Although MRI scanners are large due to the need to generate stable magnetic fields for image quality, the technique does not use ionizing radiation like X-rays, meaning 3D imaging is a powerful technique in medicine that can be used to spot cancerous tissue and study the brain.

After seeing the MRI images of the opera singer's larynx, van Andel's idea was to use the technique to create the first-ever moving images of a couple having intercourse.* This was not just an amusing, or slightly kinky, side project but an attempt to see for the first time what happens internally during coitus. After all, for some five hundred years, it was thought that the vagina was like a straight tunnel that the arrow-straight penis went in and out of. Perhaps the most famous example of this geometry is a c.1493 drawing, *The Copulation*, by the Italian polymath and painter Leonardo da Vinci.† It features a man with a straight, erect penis that is inside a woman's vagina, diagonal in direction. Fast-forward to the mid-twentieth century, and despite all sorts of efforts, including the use of artificial penises, scientists could not improve that description. Until, that is, van Andel's work and the "beautiful" MRI images taken that day.

---

* For those who want to see it with their own eyes, a video of the MRI scans can be found here: www.youtube.com/watch?v=OVAdCKaU3vY.

† Despite having little formal education, da Vinci kept extensive notebooks filled with his scientific theories, inventions, drawings, and designs. In engineering, he created once-thought impractical devices—from scuba gear to flying machines—and is particularly known in science for his work in anatomy, which includes thousands of pages of notes and beautiful anatomical drawings that fuse art with science.

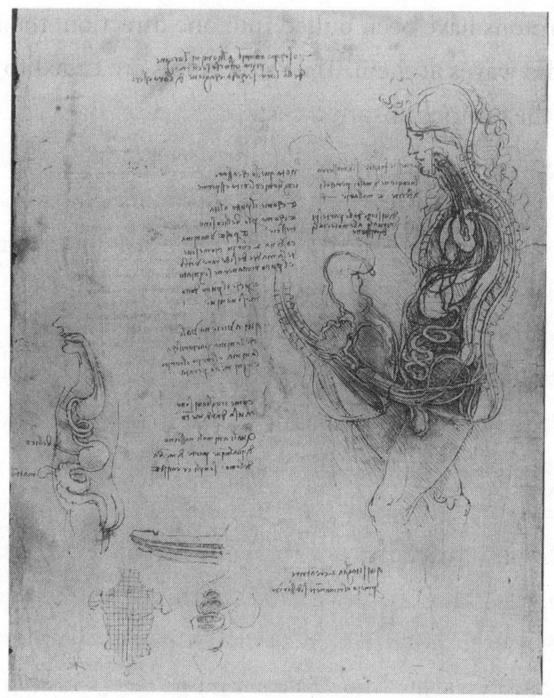

Fig 1 | Leonardo da Vinci's c.1493 drawing *The Copulation.*

After the couple's antics in the MRI machine, the team, including Sabelis, wrote up the findings and sent a scientific paper to *Nature*, one of the world's oldest and most prestigious scientific journals. It was quickly rejected, mostly due to one major flaw: a data point of one (couple). Van Andel and colleagues went back to the drawing board to recruit more volunteers to try and obtain a more general idea about what was going on during coitus while also improving the quality of the images. Getting new participants, however, turned out to be challenging. Crawling into an MRI scanner to have sex in such a small space all the while being scanned by a huge magnetic field, rather unsurprisingly, did not appeal to many.

Despite that, in 1996, the team persuaded six couples to give it a go. The team also had a new machine to play with, a Siemens Vision 1.5 tesla MRI scanner that offered sharper images. Surely nothing could go wrong? Well, it did, and the couples had performance issues. (Sabelis put her and her partner's first-time success down to their good relationship while van Andel thought it was more down to the pair's experience being street performers in their spare time.) It turned out to be harder, ahem, than it looks, with the lack of action usually pointing firmly toward the men rather than the women. "Most thought it would be easy," Sabelis, who now works at the Vrije Universiteit Amsterdam, a public university in the Netherlands, explained to me. "But they quickly realized that it takes real body awareness and a good relationship to *do* it in this way."

Two years later, a breakthrough came thanks to a new drug, Viagra. Two couples were invited back to give it another go, and with help from 25 milligrams of the drug in the form of a blue tablet, the researchers finally got the images they wanted. When the researchers analyzed the results from all the scans, they saw something quite different from what da Vinci had sketched more than five hundred years ago. Instead of a straight penis, it was curved somewhat like a boomerang. The "root" of the penis—or the first third of the shaft—was straight, but then it turned upward by an angle of around 120 degrees. This meant that the diagonal position as drawn by da Vinci was incorrect. The penis is much more vertical than was thought. The researchers also found that the size of the uterus does not increase during sexual arousal, against common thinking at the time. And they also discovered that vaginal sex in women prompts the bladder to rapidly fill, something that was observed in every participant. This happened despite each woman going to the toilet just before the scan. The reason for this is still unknown—perhaps a way to limit urinary-tract infections post coitus.

In 1999, nearly eight years after the first-ever MRI scan of coitus taking place, the results were finally published in the *British Medical Journal*,[2] one of the world's oldest medical journals. A year later, the work led to an Ig Nobel Prize for Medicine* and worldwide attention, making it to this day one of the journal's most-read articles. "It is probably my best-read publication," Sabelis said, "and likely will stay that way."

---

* The Ig Nobels are prizes that have been awarded each year since 1991 to celebrate unusual developments in science, with the stated aim of being to "honor achievements that first make people laugh, and then make them think."

THE EGG AND SPERM RACE

On a frosty day on January 2, 1665, the famous English diarist Samuel Pepys visits his bookseller. As he enters the shop in St. Paul's Churchyard in the center of London, a newly released book catches his eye. Titled *Micrographia,* the 246-page tome contains over forty beautifully detailed sketches of everyday objects such as a fly, flea, needle, piece of coal, as well as other bits and pieces. Rather than these familiar objects appearing as we see them with the naked eye, they look totally unfamiliar and strange—the drawings featuring mesmerizing aspects that had never been seen before.

The flea, for example, is described as having a "polish'd suit of sable [black] Armour" together with "multitudes of sharp pinns" that are shaped "almost like a Porcupine's Quills." As Pepys glimpses this "most excellent piece," he writes in his diary that night that the book was "so pretty" he could not resist purchasing a copy. When it arrives three weeks later, he takes it home and devours the contents in his chamber until the early hours of the morning. "The most ingenious book that ever I read in my life," he notes on January 21, 1665.[1]

The intricate drawings in *Micrographia* were made possible thanks to a new and powerful contraption—the microscope—that was about to lift the lid on this never-before-seen world. The author of *Micrographia* was the renowned English physicist Robert Hooke. He had spent several years prior to the book's publication designing and perfecting his own microscopes that used three convex lenses to produce a magnified image of an object. Hooke was a remarkable—yet controversial—scientist who made breakthroughs in several fields throughout his life, including mechanics, astronomy, and optics. At just twenty-seven years old, he was appointed curator of experiments at the then newly-founded Royal Society, the still-prominent national academy of sciences in the United Kingdom. Three years later, he published his observations in *Micrographia*, which became the world's first fully illustrated book of microscopy.

The work quickly captured the imagination and had an impact far beyond Hooke's home country. Someone equally captivated as Pepys was the Dutch businessman and scientist Antonie van Leeuwenhoek. He came from a family of beer brewers but turned his skills toward his father's side's tradition of basket making and at age sixteen began working in a textile store. Six years later, in 1654, he bought his own shop in the city. In 1668, while in his mid-thirties, van Leeuwenhoek traveled to England where it is thought he came across a copy of *Micrographia*. He was intrigued with how the microscope could help him investigate the quality of different yarns for his business. After all, *Micrographia* contained details and drawings of various threads, including silk. Inspired by the work, van Leeuwenhoek began to construct his own microscopes, which were seemingly simple but incredibly powerful instruments. One of his microscopes included a single lens—a tiny drop of glass about five millimeters in diameter (about a fifth of an inch)—mounted in a thin piece of metal, like a small magnifying glass.

Throughout his lifetime, van Leeuwenhoek created about five hundred microscopes, the best of which were capable of magnifying objects by a factor of about 250, five times the magnifying power of about fifty that others, including Hooke, could achieve at the time.[2] Van Leeuwenhoek never documented how he produced such instruments. Some say this was because of his lack of formal training as a scientist while others claim that it might have been a deliberate ploy to stop rivals from copying his techniques.[3] Yet, van

Leeuwenhoek was so ahead of the game that it took over a hundred years before lenses of a similar quality could be reproduced.

Using the most powerful of his lenses, van Leeuwenhoek could see items as small as around 2 micrometers,* which means he was able to resolve red blood cells that are between 6–8 micrometers in diameter. He was also curious enough to investigate samples from his mouth and armpits. When he studied them, he discovered something mind-boggling—they were teeming with tiny organisms that moved around, which he called "animalcules."† Then, one day in 1677, in one small ejaculation for man, one giant squirt for mankind, van Leeuwenhoek put his own semen in the spotlight. Again, he saw it teeming with "life," discovering similar animalcules in the semen of dogs, birds, and fish.‡ Remarkably, van Leeuwenhoek also measured the length of a human sperm to be around 50 micrometers long,§ resolving its head, which he found to be about 5 micrometers, or roughly a tenth of the length of the sperm. Through his research, van Leeuwenhoek not only lifted the veil on the microscopic world, as Hooke did, but also invented the field of microbiology.[4]

The 1670s marked an extraordinary decade of discovery for reproductive science,[5] with scientists finding that female mammals produce egg cells.¶ For some, the discovery of sperm cemented the theory perpetuated by the Greek polymath Aristotle in the fourth century BC that the female provided the "matter" for the baby, via the menstrual blood, while the male's

---

* A micrometer being 0.000001 meters, also written in scientific notation as $1x10^{-6}$ m.

† The original use of the term "animalcule" is thought to have been made by Johan Ham, a Leiden medical student, in 1677.

‡ It was actually Ham who was the first to see sperm, having analyzed the semen of a man suffering from gonorrhea; see: Houtzager, H.L. "Antonie van Leeuwenhoek." *European Journal of Obstetrics & Gynecology and Reproductive Biology* 15, no.3 (1983): 199–203.

§ Male fruit flies have the biggest sperm on record—measuring a whopping 6 centimeters, or twenty times their body length, when uncoiled.

¶ Despite being the largest single cell in the human body at 0.1 millimeters—possible to see with the naked eye—a human egg was not actually directly observed until 1827.

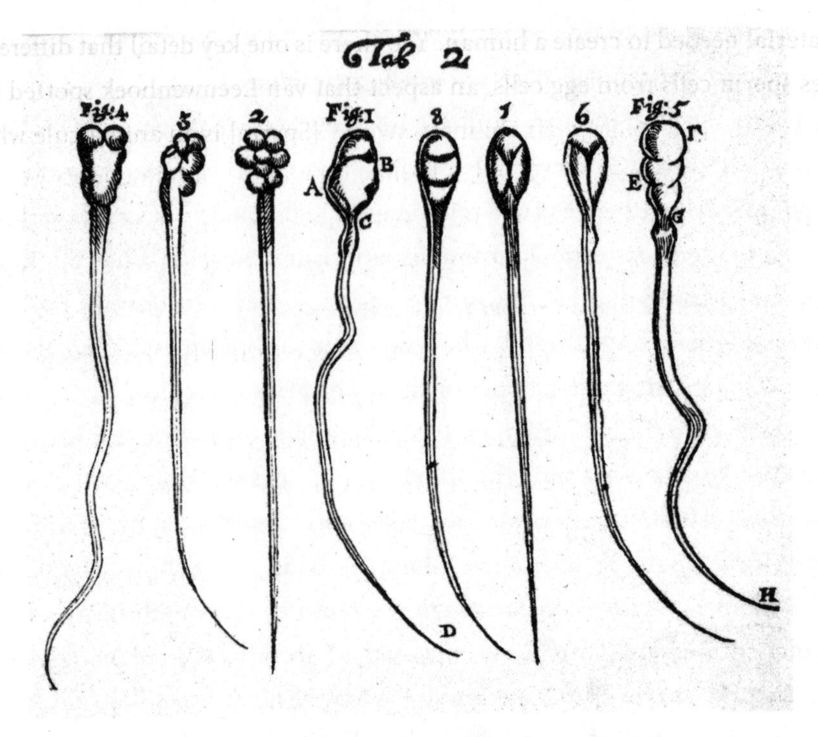

Fig 1 | *Sperm from rabbits (1–4) and dogs (5–8) as drawn by Antonie van Leeuwenhoek in the late 1670s. Wellcome Collection.*

semen gave that matter "form." This became known as the "spermist" view, while an alternative "ovist" position argued that humans were formed from eggs with the sperm or semen providing some sort of awaking force to kick-start development. Dutch microscopist Nicolaas Hartsoeker, van Leeuwenhoek's contemporary, was firmly in the spermist camp. In 1694, Hartsoeker drew a now iconic image, a homunculus, or the complete body of a tiny baby, held inside the head of a sperm waiting to pop out and grow in the uterus. Hartsoeker's sketch became a classic example of the so-called preformationist theory of human development in which people came preformed from the very beginning of conception.

This theory did not get overturned until the nineteenth century with the advent of cell biology and genetics. We now know that, despite being incredibly different in size, both the sperm and egg each contain half the genetic material needed to create a human. Yet, there is one key detail that differentiates sperm cells from egg cells, an aspect that van Leeuwenhoek spotted way back in the late-1600s: their ability to swim. "[Sperm] is an animalcule which mostly has the aspect [that] when, living and moving, it swims with its head or front part in my direction," van Leeuwenhoek wrote to the Royal Society about his first observations of sperm. "The tail, which, when swimming, it lashes like a snakelike movement, like eels in water."

While we now know that sperm must travel through the female reproductive tract to fertilize the egg, it took another 250 years after van Leeuwenhoek before anyone could offer an explanation about *how* they might do so. The first piece of the puzzle emerged in the mid-1900s, thanks to a series of experiments that highlighted just what a strange world small organisms—like sperm and eggs—inhabit.

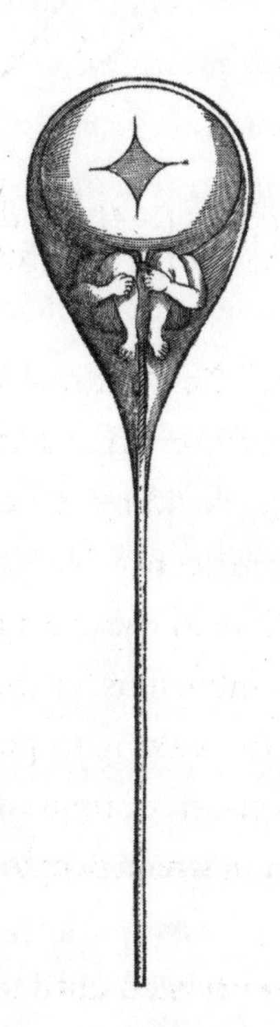

Fig 2 | *Nicolaas Hartsoeker's 1694 drawing of a homunculus. Wellcome Collection.*

❁❁❁

Human testicles are formidable sperm factories, able to pump out about 1,500 sperm cells every second,[6] resulting in 130 million cells every single day, or 10 trillion per year. In the time it took a man to read that last sentence, he would have already produced around five thousand sperm cells.* These cells travel through a bunch of tubes that make up the male reproductive system. This includes the epididymis, a crescent-shaped coil on the back of the testicle that is a remarkable 6 meters (about 19.5 feet) long in humans. Then there's a 30-centimeter-long (close to a foot long) thin-coiled structure called the vas deferens. There, sperm wait until muscular contractions during orgasm force them toward the prostrate, where they are washed together with semen and then pushed straight out of the penis via the urethra.

It takes around three months from sperm-cell production to full maturation. An average ejaculate contains around fifty to one hundred million sperm—in pure numbers enough to produce the population of a country.† Why men produce so much sperm remains a mystery, but it could just be a numbers game.‡ The ejaculate lands at the top of the vagina (which is about 7 centimeters, or 2.8 inches, long), and for around 95 percent of sperm, this marks the end of the road for several reasons. The first is that sperm are exposed to the slightly acidic fluid of the vagina.[7] The second and somewhat bigger issue is the makeup of the sperm themselves. A large percentage, as much as 90 percent, are deformed. Some have bent necks or misshaped heads,

---

* Although these aren't fixed numbers—counts can vary substantially person to person.

† Again, counts can vary a lot, and this number is nothing compared to a ram, which can produce around ninety-five billion sperm in a single pop. See www.open.edu/openlearn/nature-environment/natural-history/sperm-counts.

‡ Another reproductive strategy is producing a few but large sperm. In 2020, scientists discovered the oldest sperm ever seen in a previously unknown species of crustacean. The sperm, produced around one hundred million years ago, was several times the length of the ostracod's body; see: Wang, H; Matzke-Karasz, R; Horne, D.J; et al. "Exceptional Preservation of Reproductive Organs and Giant Sperm in Cretaceous Ostracods." *Proceedings of the Royal Society B* 287 (2020): 20201661.

and some don't even have heads.[8] Of the 10 percent that are "normal," around half cannot swim very well, going around in circles or doing nothing at all. They just wither away. From starting with one hundred million sperm, the ejaculate is already down to five million—not a particularly good start.

The sperm that can move begin to maneuver themselves through the cervix, a narrow channel filled with mucus that is around 2 centimeters (about three-quarters of an inch) long. Cervix, Latin for "neck," is a gatekeeper, letting some things in and some things out. The sperm continue their obstacle course as they enter the uterus, which is about 8 centimeters (about 3 inches) long and shaped like an upside-down pear. At the top of the uterus on both sides are narrow fallopian tubes that are around 7 centimeters (about 2.8 inches) long. At the end is the ovary itself. All these lengths seem small, but given the tiny size of the sperm as measured by van Leeuwenhoek's microscope, it is a monstrous total distance to travel—equivalent to a human swimming a hundred lengths of an Olympic-sized swimming pool.

Sperm are not alone in their journey. They arrive at the top of the vagina encased in semen, a cloudy-white fluid with a jellylike consistency. As the vagina's environment is slightly acidic, one of the ways semen protects the sperm in this inhospitable setting is by increasing the pH in the vagina from

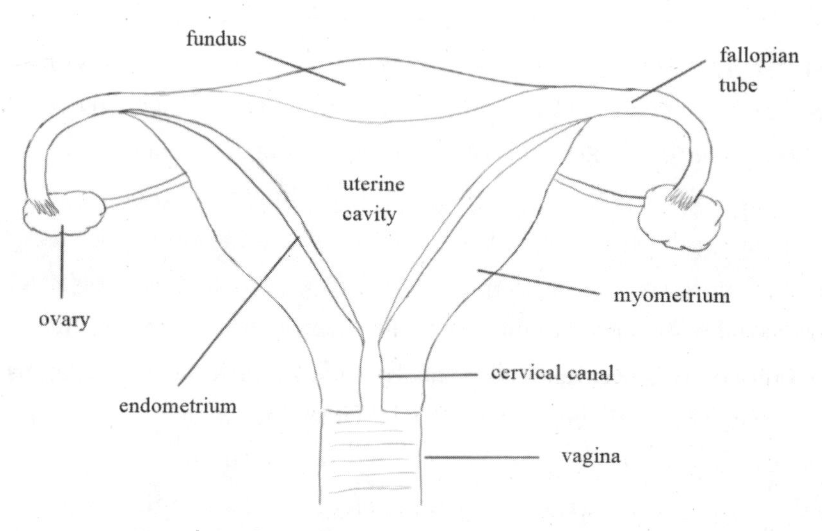

Fig 3 | *Outline of the female reproductive system.*

5 to 7. The sperm then enter the cervical mucus, which has a similar egg-white-like consistency, before traveling on to the uterus that is also thought to consist of watery mucus.[9] The final part of the journey is then through the fallopian tubes. As sperm spend all their time in fluids such as semen or the cervical mucus, they need a way to propel themselves in such substances. But it turns out to be incredibly tricky for such a small cell to move, so to understand how they manage to swim through these fluids, we need a crash course in fluid dynamics—the study of the flow of liquids and gases and their interactions with solid surfaces.

A fundamental aspect of fluids is their viscosity, which is defined as the resistance of a fluid to a change in shape or movement. A fluid with high viscosity, like honey, resists motion because its molecular makeup produces a lot of internal friction. A fluid with low viscosity, like water, flows easily because its molecular makeup results in very little friction when it is in motion. As an example, imagine a cup with a small hole in the bottom. If honey or oil is poured into the cup, it will drain slowly because of the high viscosity. Water, on the other hand, drains much more quickly due to its lower viscosity.

A way to describe the properties of different fluids by their flow and how objects can move in them came in the late 1880s from the Irish physicist Osborne Reynolds, who was a professor of engineering at Owens College in England (later reconstituted as the University of Manchester). During the 1870s and 1880s, Reynolds conducted a series of experiments in which he injected colored dye into a small section of a thin pipe that contained water. By changing the speed of the water, Reynolds could test under what conditions the flow would be smooth or turbulent. With incredible insight, Reynolds found one simple number that could describe the balance of the forces involved for an object in a fluid—the Reynolds number, known simply as R_e. It is defined as a ratio between inertia—the tendency of matter not to change velocity—and viscosity.[10] The inertial force depends on the size and velocity of the swimming object while the viscous force depends on the density of the fluid. Loosely speaking, a Reynolds number greater than one means that inertial forces are dominant; less than one means viscous forces rule.

The Reynolds number would later become important in engineering from designing airplane wings to tweaking the aerodynamics of Formula 1 cars (air is considered to be a moving fluid), but it also plays a huge role in biology, which can throw up an enormous range of values. A whale, for example, has a Reynolds number of about one million while a swimming human's is about ten thousand.[11] What a large Reynolds number tells us is that for humans or a bigger animal like a whale, the inertial forces created by the moving body dominate over the viscous, or resistive, forces of the water. Indeed, a flip of a whale's tail allows it to travel large distances with little resistance from the water even for such a large body. Things are completely different for microorganisms such as bacteria and sperm. They tend to have a Reynolds number that is much smaller, tiny in fact, at around 0.0001. In this case, it is not inertial forces that are at play, but rather viscous forces.

An elegant way to show just how hard it is for microorganisms to swim was made a hundred years after Reynolds by the American physicist Edward Mills Purcell. He is best known for the discovery of nuclear magnetic resonance in the 1940s, a breakthrough that paved the way for MRI, which was introduced in chapter one.* Purcell also had a penchant for back-of-the-envelope calculations, and in the 1970s, he became interested in what he called the "majestic" swimming of microorganisms.[12] In 1976, Purcell gave a now-famous lecture in which he outlined just how hard it is for bacteria to move in a liquid. The physicist calculated that if you gave a bacterium a tiny push in a liquid it would stop within a millionth of a second.[13] In that time, it would have traveled a distance less than the width of a single atom. Purcell highlighted that bacteria inhabit a world where inertial forces are completely irrelevant—one that is vastly different from what we are accustomed to. For a human to emulate

* Purcell shared the 1952 Nobel Prize in Physics with the Swiss-American physicist Felix Bloch for their work on NMR.

how difficult it is for a microorganism to move, we would have to try to swim in a medium with a high viscosity like honey *and* move our arms at the same speed as the minute hand of a clock. If it were actually possible to simulate this, it would take weeks to travel a few meters. Exhausting.

The physics explaining how all this happened was worked out about two decades prior to Purcell, notably by several physicists in the United Kingdom, including Geoffrey Taylor from Cambridge University. In a series of classic experiments in the 1960s using glycerin, a high-viscosity medium, he showed just how bizarre this world is.* At a low Reynolds number, the physics of a swimming microorganism is all about the ability to break reciprocal motion, a repetitive movement such as up and down or side to side that prevents locomotion in viscous fluids. The simplest example of reciprocal motion, as elucidated by Purcell, is the humble scallop. If you shrink a scallop to have a low Reynolds number like sperm or bacteria, the scallop would not be able to move.† This is because its movement is totally reciprocal. When the scallop opens and closes its shell, it goes through the same motion for the power stroke (closing the shell) and the recovery stroke (opening the shell). Another way to think about it is to film the scallop closing and opening its shell. If you played this video forward or in reverse, you would not be able to tell which stroke was which. Basically, the micro-scallop is trapped in time.

Yet, we know that microorganisms can swim. After all, if they could not, you would not be reading this and I would not be around to write it. So how can they do it? Again, Taylor showed how this could be done. If you take a thin cylinder like a straw and let it drop upright in a high-viscosity fluid like syrup, it will fall vertically, as expected (see figure). If you put the straw on its side, it will still drop vertically, but half as fast as in the upright case due to increased drag. However, when you put the straw at an angle from the horizontal position—think like the back of an inclined seat—then it does not just

* For a video of this in action, see: youtu.be/53Rgjmr2rl0.

† Actual (normal-sized) scallops move by jet propulsion, squeezing the fluid as they close the shell.

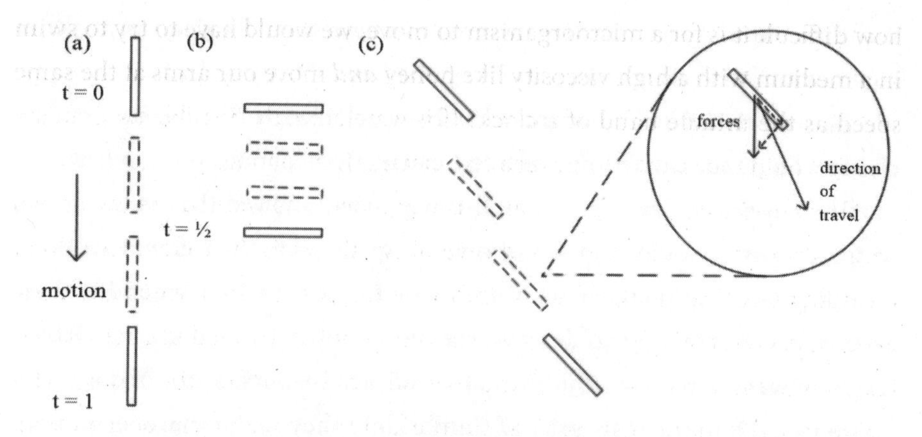

Fig 4 | *When a rod falls vertically upright in a highly vicious liquid, it moves straight down (a), and when placed horizontally (b), it moves half as fast as the upright version due to increased drag. When placed at an angle (c), however, the rod moves diagonally, known as "oblique motion," due to how the forces (inset) act on the body.*

move down the syrup vertically but also travels horizontally, resulting in it falling in a diagonal direction.[14] This is known as "oblique motion," and why it happens has to do with how the forces act on the slim body. The vertical force in this orientation can be given as two components, one along the length of the body and one perpendicular to it (see figure). The lower drag along the length of the body results in a greater motion in that direction compared to the perpendicular, meaning that the straw wants to move along its length faster than it does perpendicularly, so it slips horizontally as well as falling vertically.

You may well ask what this has to do with swimming bacteria and sperm. Well, again, they must break reciprocal motion to move, and Taylor showed a particular way they could do so. The most basic way—and one found countless times in nature—involves the spiral rotation of the tail, or flagellum, that sticks out from the main cell body. This tail moves like a rigid corkscrew (imagine opening a bottle of wine, much needed after a day with kids), and this helical rotation is what breaks reciprocal motion for low-Reynolds swimmers. Imagine splitting the helix up into smaller segments and then deducing the amount of oblique motion in each part, which is then summed up to estimate a forward propulsion. Indeed, this helical trick is employed by bacteria such as *E. coli*. These effective swimmers rotate their flagella clockwise or counterclockwise via motors that are on the base of the flagella.[15]

In the early 1950s, Taylor and Geoff Hancock from the University of Manchester, England, carried out detailed calculations about how a cell with a moving flagellum, like sperm, could travel.* They showed that as the sperm whips its tail, it could also create oblique movements at different sections, creating viscous propulsion as it does so.[16] In 1955, Hancock applied these mathematical principles to describe the motion of sea urchin sperm.[17] At the time, he was working at Queen Mary College, University of London, with James Gray from the University of Cambridge. They found that sperm break the time reversibility by using the elasticity of their tails to perform complex wavelike "beating" movements that create oblique propulsion.

To carry out these movements, the sperm's tail, as with all flagella in nature (which will also come into play in the next chapter), needs some biological machinery. And, just as van Leeuwenhoek used newly created lenses in the 1600s to view individual sperm cells, researchers in the late 1950s used beams of electrons from a transmission electron microscope (TEM) to delve even deeper into the sperm tail's construction.[18]

TEM was invented in 1933 by German scientists Max Knoll and Ernst Ruska,† and using the new device the pair discovered a beautifully intricate and, in some sense, simple structure. A sperm's tail contains a fibrous sheath within which are clumps of dense fibers arranged in a circle. The middle of this circle, called the axoneme or cytoskeleton, is where the sperm gets its powerful beat. The main components of the axoneme are long tubes called microtubules. The axoneme features nine pairs of these microtubules in a ring with a single pair of microtubules in the middle—dubbed a "9 + 2" arrangement—that runs down most of the tail, apart from a few micrometers right at the end

* At the time, Hancock was a PhD student, working under the guidance of Sir Michael James Lighthill, a pioneer in the field of aeroacoustics and biological fluid dynamics.

† Ruska was later awarded the 1986 Nobel Prize in Physics for the discovery. Knoll died in 1969 and would likely have been awarded the Nobel Prize along with Ruska, if it were not for the fact that Nobel Prizes are not awarded posthumously.

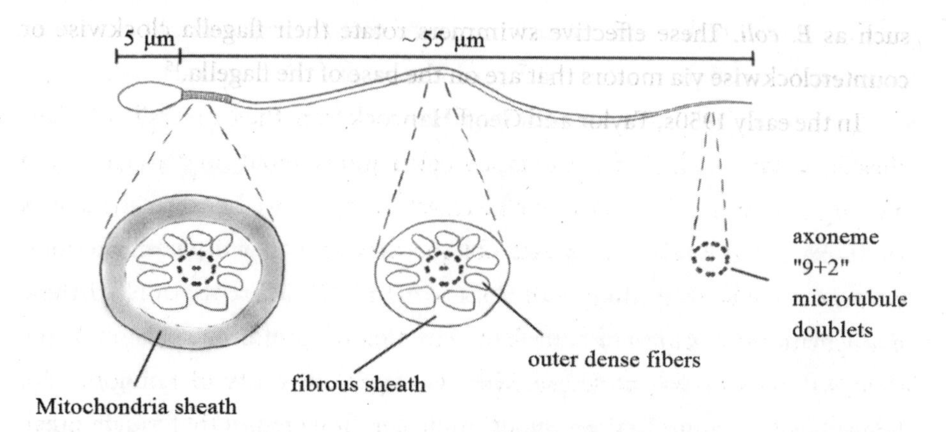

Fig 5 | *The main component of a sperm's tail is the axoneme, which contains sets of microtubules arranged in a "9 + 2" arrangement. The axoneme toward the front of the tail is enclosed by outer dense fibers and a fibrous sheath.*

(see figure).[19] Motor proteins called dyneins, responsible for connecting the pairs, enable the microtubules to slide past each other, which results in the bending of the whole tail. Pure biomachinery in action. A sperm's tail can even "counterbend." Push the tail at one end in one direction and the other end moves the other way (even dead sperm can counterbend[20]).

But there is more to it than just complex biomechanics. Sperm must still swim in the cervical mucus, which changes consistency, or viscosity, throughout the menstrual cycle, especially around ovulation. For most of the menstrual cycle, the fluid is thick and dense like toothpaste, making it impregnable to sperm. But near ovulation the composition changes thanks to the release of estrogen, which is a female sex hormone produced by the ovaries, as well as the placenta during pregnancy, and is responsible, among other aspects, for helping the uterus grow and aiding the development of the breasts to prepare them to produce milk once the baby is born. The cervical mucus then becomes similar to egg whites (uncooked, of course): clear, profuse, and slippery.* Even at this point, the cervical mucus can have a viscosity up to two

---

* If you put the watery, egg-white-like mucus on a slide, let it dry, and then view it with a microscope, you will see a pattern that resembles a fern tree. This is known as the "fern test" and is an important diagnostic tool to deduce ovulation (see chapter four).

hundred times greater than water. Despite the thought that it would be harder for sperm to swim in this gloopy, egg-white-like fluid, it turns out that, rather strangely, it can be somewhat advantageous.

"If you want to know how sperm swim, you are in the right place," Hermes Bloomfield-Gadêlha, a Brazilian-born mathematical biologist at the University of Bristol in the United Kingdom, tells me as I meet him at his office on a late autumn day. Bloomfield-Gadêlha has spent his career working on the mathematics of swimming sperm, bringing together the field of fluid dynamics with the molecular machinery of the sperm's tail. But what really shines through is Bloomfield-Gadêlha's enthusiasm, so much so that he is late for his next appointment, lost in the low-Reynolds world. Before then, Bloomfield-Gadêlha shows me a film of a sperm swimming in a liquid with a consistency of water. Here the sperm whips its tail in all directions, up, down, left, and right. The tail "beats" with a frequency of around 25 hertz* (equivalent to 25 cycles per second) and seems to roll around as it swims.

The film gives the impression that the movement of the sperm through the liquid is random, even chaotic, but previous work has shown that something surprisingly different is happening with the fluid. The team recorded a sperm swimming in a saline solution and then extracted its movement to simulate the resulting flow of the fluid—somewhat like putting a big rock with moving parts in the middle of a river to see how its movements change the flow of the water.† They found that the fluid flow around a swimming sperm in

---

Ferning can be seen from the sixth to the twenty-second day of the menstrual cycle but becomes most prominent at around the twelfth to sixteenth days—or when a woman is most fertile around ovulation.

* Hertz is the unit of frequency, which is the inverse of the number of cycles per second.

† The mathematics involved to describe the motion of Newtonian fluids is the Navier-Stokes equations—a set of partial differential equations. A form of these equations to describe low-Reynolds swimmers like sperm in a microscopic environment is called the Stokes equations, and the mathematical abstraction of a single point force in the "Stokes" creeping flow that results is known as a "Stokeslet."

a low-viscosity fluid like water follows a well-defined, smooth pattern despite seemingly splashing around. If you cast your mind back to high school physics class, you may remember the classic experiment with magnets and iron filings. Place a bar magnet underneath a piece of paper and then scatter iron filings on top. The iron will become magnetized and align along the bar magnet's magnetic field lines. In this case, as a sperm swims, it stirs the fluid, producing similar field lines in the fluid. In this sense, the swimming sperm is like the bar magnet creating a dynamic "field" in the surrounding fluid.[21]

That is all well and good for water, but we know that sperm need to move in the high-viscosity fluid of the human cervix. And, it seems, sperm are engineered to do so. Bloomfield-Gadêlha next shows me a video of sperm swimming in a high-viscosity liquid, and the swimming behavior is completely different, mesmerizing in its simplicity. Here, the sperm's head is much more stationary and only the tail seems to move—looking like a slithering eel, as van Leeuwenhoek first described. The difference between low and high viscosity is almost like one person flapping their arms when learning to swim and the other doing smooth breast strokes. "The sperm is in a completely different gear when in mucus," Bloomfield-Gadêlha explains.

The mathematical models of swimming sperm that have been developed since the 1950s have enabled researchers to play around with certain aspects of the sperm's tail to see what aspects are important. Bloomfield-Gadêlha and colleagues have examined the role that the outer dense fibers play. The sheath surrounds only the top of the sperm's tail, extending around a third of the way down, which gives the top part of the sperm's tail more stiffness than the middle part. The team studied the sperm of sea urchins, which swim in the low-viscosity medium of seawater to fertilize the egg. But when they are placed in a high-viscosity medium, they swim much more slowly. When Bloomfield-Gadêlha and colleagues modeled an outer sheath onto the sea urchin sperm, they remarkably found that sperm could move more effectively in a high-viscosity fluid.[22] This showed that the reinforcing outer layer, which coats the tails of human sperm, plays a key role in making powerful rhythmic strokes that guide it through high-viscosity fluids. "We don't know what came first, the sheath or the cervical mucus. Perhaps they co-evolved?" says

Bloomfield-Gadêlha, who is now developing techniques to watch sperm swim for minutes rather than just seconds as in current techniques.[23] "But nothing in nature happens by chance."

Out of millions of sperm in an ejaculate, only a few hundred reach the fallopian tubes, also known as the oviducts, located near the top of the uterus.[24] At the end of the fallopian tubes are the ovaries, and somewhere in the middle section—called the ampulla—will be the egg. The lucky sperm that reach the egg can now sense it, which sets off a whole new pattern of movement—and one that is much more chaotic than the smooth swimming in a high-viscosity medium.

Once the sperm approach the egg, perhaps a couple of millimeters (around a tenth of an inch) away, they detect the hormone progesterone, which is released by the egg, and move toward it in a process called chemotaxis, in which the movement of cells and organisms are directed by the chemicals in their environment. Progesterone is contained in the follicular fluid, a nutrient-rich fluid that surrounds an egg. As the egg develops, the hormone lures the sperm toward it. Intriguing research in 2020 found that the follicular fluid can attract more sperm from some men than others—an effect that seems to be random and did not correlate with a woman's chosen partner.[25]

The mechanics behind this powerful effect on sperm is a calcium channel called CatSper (which stands for cation channel of sperm) that was discovered along the tail of human sperm in 2001.[26] This CatSper protein receives progesterone and sends calcium into the cell, which causes the sperm to go into a frenzy—called hyperactivation. Here, the smooth swimming action along the mucus of the cervix is now replaced with a chaotic whipping of their tail. While this may give the impression that the sperm has no chance of getting anywhere, the motion gives it two distinct advantages. The first is that it stops the sperm from getting stuck in the fallopian tubes. The second is that it causes the head to switch from a sideways motion to a figure-eight twisting motion. This hammering-like movement makes it ideal for boring through the egg's zona pellucida, a jellylike protective layer that is between 13–19 micrometers thick, or about two or three lengths of a sperm's head.

Experiments in 2020 showed that the CatSper protein is incredibly important. Without it, the sperm were unable to fertilize the egg.[27] Yet, the sheer force of the sperm's drilling is still not quite enough to break the barrier. To quicken things up, sperm release a series of enzymes, or proteins that speed up reactions, that are in the acrosome, which is at the tip of a sperm's head. This helps to dissolve the zona pellucida, resulting in a bit of an onslaught on the egg—hammering and dissolving at the same time.

The interest in how sperm swim in either low- or high-viscosity fluid is not purely academic, and mathematicians are teaming up with reproductive specialists to investigate if all this knowledge of the mathematics of swimming sperm can improve the diagnosis for couples undergoing fertility treatment. Across Europe and the United States, one in six couples is infertile, and each year the number of people referred to infertility centers increases by about 9 percent. In the United Kingdom, for example, over fifty thousand women undergo fertility treatment each year—almost double that seen in the last two decades—resulting in more than seventy thousand treatment cycles.[28] A major reason for this is that sperm counts in men have declined by half over the past forty years, with one in twenty men now having a low sperm count. It is estimated that around one hundred million men worldwide may be "subfertile," which has led to warnings of a "fertility time bomb." Male-factor infertility and unexplained infertility are now the dominant reasons why couples are turning to assisted-reproduction techniques.[29]

On a cold, wet day in February 2020, I meet mathematician David Smith from the University of Birmingham in the United Kingdom. We head to Birmingham Women's Hospital, which is only one of two dedicated women's hospitals in the nation and is located directly opposite the newly constructed and impressive-looking Queen Elizabeth Hospital Birmingham—one of the largest single-site hospitals in the UK. Birmingham Women's Hospital itself features the busiest single-site maternity unit in the country, delivering over eight thousand babies a year, and is also home to a leading fertility center.

We enter by the side entrance, the place where people drop off their sperm samples for analysis or donor submissions. We walk past the inpatient reception area and head to the third floor to meet Jackson Kirkman-Brown, who is waiting for us outside his office.

Wearing a trademark waistcoat and bow tie, Kirkman-Brown is a world-renowned fertility expert, particularly when it comes to male fertility issues. After earning a PhD at Birmingham University studying the interaction between human sperm and eggs, he went to the University of Massachusetts Chan Medical School for a year before heading back to Birmingham University. During the Iraq War (2003–2011), Kirkman-Brown was involved in helping to extract sperm from severely injured soldiers, allowing them the chance to still have children. In 2013, he was recognized for his work in the Queen's New Year's Honors, and awarded a Member of the Order of the British Empire (MBE) for services to human reproductive science. Kirkman-Brown now spends half his time at the university and the other half as the science lead at the fertility center, where he specializes in male-factor fertility.

Kirkman-Brown tells me about experiments with rabbits in the 1970s, led by Jack Cohen from Birmingham University, in which researchers recovered the small population of sperm that reached the uterus and oviduct and then re-inseminated them into another female rabbit of a different breed together with the full ejaculate of another male. By keeping track of certain characteristics, such as coat color and pattern, they showed that this tiny population of recovered sperm could make the journey twice.[30] Although the conclusions are not universally accepted, the studies point to the idea that there are certain groups of sperm that have advantageous characteristics, which set them apart from the rest. "Only tens of sperm arrive at the egg, and we still don't know what the characteristics of this set of sperm are," Kirkman-Brown says. "Just looking at all the sperm in a semen sample won't tell you that information."

When couples begin infertility investigations, it often involves the analysis of a semen sample by trained technicians. This analysis gives an indication of various parameters, such as ejaculate volume, sperm count, motility, and morphology. While this approach is used in many reproductive centers to determine the level of infertility, the accuracy of the technique is not guaranteed. It is also expensive and time consuming. The current "gold standard" technique is computer-aided sperm analysis, or CASA, in which a sample is loaded into the microscope and then a computer takes images of the sample for about one second, counts the number of sperm, and looks at certain characteristics. Yet, both manual methods and CASA tend to focus on the sperm's head when analyzing the sperm sample, not only to identify sperm in the first place but also to monitor their swimming ability.[31] While some sperm may have the correct head morphology, and even good motility, they may be hiding defects that would make it impossible to ever reach the egg. "It's really the tail that gives you a metabolic readout of the cell," Kirkman-Brown tells me over a coffee in a busy employee common room.

Smith, who has worked on the mathematics of swimming sperm for the past two decades, is collaborating with Kirkman-Brown and other reproductive specialists in Birmingham to investigate new mathematical methods to analyze sperm for reproductive medicine. The team has developed a new sperm-analysis technique—dubbed Flagella Analysis and Sperm Tracking (FAST)—that can capture and analyze the tail of a sperm in exquisite detail. It uses rapid digital-camera imaging to quickly take several images of a sperm as it swims in a saltwater-like liquid. Measuring the sperm's "waveform," the researchers can pull out many characteristics of the swimming cell, such as the frequency of the tail beat (usually about 25 hertz for healthy sperm) as well as the swimming speed (about 50 micrometers per second).[32] The program uses the mathematical techniques developed by Smith and colleagues to model this movement and calculate how much force the cell is applying to the fluid and its swimming efficiency—how far it moves using a certain amount of energy. All this information could give an indication of whether the sperm would ever have the capacity to reach the egg and fertilize it.

The benefit of FAST is that it can assess large quantities—up to hundreds of sperm at a time—and then, if necessary, extrapolate the results to the whole sample. The team has begun clinical trials with FAST involving seventy-three couples and around 14,000 individual sperm, with plans to expand it further.

FAST could also be used to study the effect that lifestyle or supplements have on sperm motility and even, perhaps, to calculate the effectiveness of male contraceptives. But, ultimately, this technique is to determine whether fertility treatment is necessary in the first place, as undergoing assisted reproduction can sometimes feel like taking a hammer to crack a nut.

If there were a better method to analyze sperm samples, for example, then it might be possible to use an assisted reproduction technique that is perhaps less invasive and, therefore, cheaper. The simulations may show that intrauterine insemination could be just as successful over several cycles as carrying out IVF if the issue is suspected male-factor infertility. This insemination approach involves putting the washed and concentrated sperm into a syringe and directly squirting them up into the uterus, bypassing the cervical canal. According to Smith, even if in the future just 5 percent of those seventy thousand rounds of assisted reproduction each year in the United Kingdom could be avoided thanks to a better analysis of sperm, it could save over $1 million annually—all thanks to an understanding of the mathematics of how sperm swim.

What decades of research has shown is that swimming is tough for small cells like sperm. But thanks to the fascinating mechanics of their ingenious tails, as well as the properties of the cervical mucus, it is possible for sperm to swim to the egg. Yet, that is not the whole story. A common misconception with conception is that it is the sperm that does all the work while the egg, which is about 0.1 millimeter in diameter (roughly about the size of the period at the end of this sentence), lies dormant waiting to be fertilized.* The female reproductive system itself can pull several levers to aid sperm on their journey, one

* While men continuously produce sperm throughout their lifetime, women are born with a finite number of eggs—around a million, but by the time of puberty this has reduced to about three hundred thousand. During each menstrual cycle, an egg is matured in a follicle in the ovaries before erupting from it and is released together with a cloud of cells that surround the egg.

of which is muscular contractions in the uterus (which will be critical when it comes to birth, as we will see in chapter six) that drives uterine fluid to the fundus, or the top of the uterus.

During menstruation, it is thought that contractions begin from the top of the uterus and move down toward the cervix at a rate of about one contraction per minute to help expel the lining of the uterus. During the remainder of the menstrual cycle, however, the direction is not only opposite, but the contractions are quicker, around three per minute.[33] Combining this muscular boost with the sperm's swimming ability results in the sperm reaching the fallopian tubes in less than twenty minutes—an incredibly short time given their microscopic size. Once there, it is difficult to know exactly what is happening in vivo in the fallopian tubes, given their complex, labyrinth-like structure. But it is generally thought that sperm can be stored for days in special "crypts." It has even been suggested that sperm are somehow released in batches. In effect, the female reproductive system controls the volume of sperm progressing toward the egg. There may be a reason for this. Men who have high sperm count have an increased risk, although it is still exceedingly rare, that two sperm might simultaneously fertilize an egg. If this happens, the embryo will contain sixty-nine chromosomes instead of forty-six, resulting in a miscarriage or early death if the baby gets as far as being born. These crypts could be a way to lower the probability that this will happen, although given the challenges of experimentally testing this, it is far from clear whether this is the case.

What we do know is that out of the millions of sperm that are in an ejaculate, only one, possibly, will enter the egg. It is a chance of success akin to winning the lottery but with a reward that is incalculable—life itself. However, much research is still needed to fully understand the precise details of how microscopic sperm reach the egg, and whatever future surprises emerge, one aspect will always remain:

Life begins at a low Reynolds number.

3

FORMING A BODY PLAN

Nature is full of patterns. Think of the perfect, replicating arrangements of flower petals, the intricate spiral patterns of permanent marker drawn on the wall by a toddler, or the repeating stripes on a zebra. Even hold out your arm, and you will see five nearly-identically formed fingers on both hands.*

The British mathematician Alan Turing was fascinated by how all these different patterns could form in the natural world. He was not a trained biologist, instead studying mathematics at Cambridge University before heading to Princeton University in the mid-1930s for a PhD in mathematical logic. Turing rose to fame, however, when he returned to England before the onset of the Second World War to join the code-breaking activities at Bletchley Park, where many men and women worked day and night to decipher the communication codes that were produced by the German navy's Enigma machine to encrypt radio communications.

After those heroic code-breaking efforts, Turing continued his passion for computing and artificial intelligence. He also turned his attention to one of his

* About one in five hundred people are born with polydactyly—extra fingers or toes.

other interests: embryology. Turing was inspired to do so by the work of the Scottish biomathematician D'Arcy Wentworth Thompson, who, in 1917 while at University College, Dundee, published his now-classic book *On Growth and Form*.[1] The tome investigated many different topics, from the development of trees to bone structure and skeletal dynamics. It also contained a mathematical description of morphogenesis, the biological process that gives living things their shape. Despite some of his ideas facing opposition at the time, D'Arcy was convinced that mathematics could explain the formation of living organisms such as animals and plants.[2] Turing was equally certain. Just as a computer follows a set of instructions in code, this same underlying logic must also apply to biological organisms.

Building on the ideas of Thompson, Turing focused on how morphogens—substances whose distribution governs the pattern of tissue development—could diffuse in a defined space. In other words, from a purely theoretical perspective, he looked for the unfolding mechanisms that are responsible for a wide range of patterns we see in nature.

The main thrust of Turing's theory involved two competing actions diffusing through a space, such as in a developing embryo. One is an "activator" that switches something on, say a pigment, while the other is an inhibitor that stops that process from happening. A key aspect of Turing's theory is that when the activator moves, it not only replicates itself but also makes the inhibitor too. Another feature is that the inhibitor moves at a greater speed than the activator.

Fig 1 | *Schematic of various Turing patterns.*

The theory was simple and elegant, but it nevertheless came with some mean-looking mathematics. In 1948, Turing moved to the University of Manchester, where a new computer—the Ferranti Mark 1—had just arrived that could help him carry out simulations that would visually show his mathematical theory in action. Crunching through the equations, Turing showed that if you start off with a little bit of activator and set it on its way, by tweaking the speed of propagation of both the activator and inhibitor, it was possible to create a myriad of different patterns from simple spots and stripes to even more complex arrangements, such as rings, dappled blobs, or a labyrinth pattern.[3] The geometry of the diffusion also created its own possibilities—for example, a spot-like pattern on a cheetah could appear as stripes on a thinner section, such as a tail, just as seemingly happens on the animal itself.

Turing's idea was a classic example of how physicists and mathematicians typically approach a problem—strip it down to its most basic components.* As a result, he did not apply the theory to a particular problem in biology and acknowledged that the theory had many shortcomings. "The theory does not make any new hypotheses," he wrote in his 1952 paper. "It merely suggests that certain well-known physical laws are sufficient to account for many of the facts." Sadly, Turing would be unable to develop it further. Following his arrest for "gross indecency" after having a relationship with another man (homosexuality was illegal in the United Kingdom in the early 1950s and Turing was openly gay), Turing was given the penalty of chemical castration. In 1954, just two years after his theory was published, he took his own life at age forty-one.†

Others did take on the mantle. In 1988, the Scottish applied-mathematician James Murray, who was then working at the University of Washington in

---

* The expression "consider a spherical cow" comes to mind when physicists model living things.

† Some spent years campaigning for Turing to be given a Royal Pardon, which was eventually granted by Queen Elizabeth II in 2013.

Seattle, took Turing's idea further and used computer modeling to show that the reaction-diffusion theory could explain most of the animal coatings found in nature.[4]

Despite the genius of Turing's idea, however, it took biologists decades to see its possibilities. Even then, many were—and still are—skeptical that a relatively simple theory could explain such complex results. Many developmental biologists were instead taken by a model in the late 1960s devised by the British embryologist Lewis Wolpert who worked at University College London. This theory dictated that cells could sense where they were in relation to an underlying map of molecular signals in the embryo, which would result in the creation of different structures. It is somewhat akin to a children's coloring book that has a picture in outline and then numbers in certain spaces to represent different colors. The outline is the underlying map, and then as a morphogen spreads, it deposits the correct "color" in each area to make up the whole multicolored picture.

Wolpert's model has had a lot of success in real-life systems. One example is in the generation of the body of the fruit fly (*Drosophila melanogaster*). The fly is a model organism to study for biologists given that many of the genes that control the development of flies are like the ones that control vertebrates. The body of the fruit fly is not generated by the reaction-diffusion mechanism, but rather the varying concentration of morphogens—in particular, the protein bicoid—as they diffuse across the body.

But Turing's model was not totally down and out. In 2006, biologists in Switzerland found definitive evidence for Turing's theory in the placement of hair follicles.[5] It also turned out that digits on a hand, or a paw, can be a perfect demonstration of Turing patterns. In 2012, biologist James Sharpe from the Centre for Genomic Regulation in Barcelona, Spain, and his colleagues proved the effectiveness of Turing's method in producing digits in mice. They removed a set of genes called Hox, which is part of a large family of genes that are involved in many areas of development, including organizing a body plan. As they knocked out thirty-nine Hox genes one by one, the mice had more and more fingers, going up to a maximum of fifteen.[6] This genetic change did not result in the mice having mutant-sized paws. Rather, they were the same size, but the spacing between the digits was reduced.

The study was a classic example of a Turing pattern in which spots turn into stripes when crammed into a smaller space, and calculations carried out by the researchers based on Turing's equations replicated the pattern they saw in the experiments. Two years later, Sharpe and his colleagues took the idea further and found that the complex interplay between three different molecules could explain why fingers grow where they do. They found that a certain protein, dubbed SOX9, signals to build bone at a specific place, while another protein switches on SOX9 and another turns it off in the gaps between the fingers.[7]

It is plausible that nature uses a combination of reaction-diffusion and morphogen gradient mechanisms to create a body plan. Yet Turing patterns alone can still explain a range of phenomena from the creation of branches in the lung to the ridges on the top of your mouth—the ones that get burned when you bite into a nuclear-hot Pop-Tart. Turing's insights have also found applications beyond embryo development, such as explaining aspects of galaxy formation and predator-prey relationships in ecology, and even shining a light on crime hot spots in cities. But what Thompson and later Turing and others kicked off was a rigorous mathematical and physical approach to understanding problems in biology, specifically embryogenesis and morphogenesis. And, while biologists have for decades focused on how genes shape us, there is now a growing body of evidence that this alone is not enough. A deeper knowledge of the physical forces in embryos is raising answers and more questions about how we come to be.

When two compatible people meet, we often say that sparks fly, and that is exactly what happens when an egg and sperm meet—or at least the zinc sparks fly. In 2016, scientists from Northwestern University in Chicago used high-speed cameras to film the "fireworks" that occur when a human sperm and egg come together. They found that when a sperm enters an egg, it leads to a surge of calcium in the egg that triggers the release of zinc. As the zinc shoots out, the researchers made it bind to small molecules that then emit light so the zinc could be detected. They found that the egg distributes zinc to control the development of a healthy embryo. The more zinc that is released when this happens, the brighter the flash of light, and potentially the more viable the transition is from an embryo to a fetus.[8]

When that spark happens, the merged genetic material produces the first cell, called the zygote, which starts to divide in a process called mitosis. This first cell division occurs around twenty-four hours post fertilization, and by the end of the fourth day, the zygote already contains around sixteen cells. The embryo travels down the fallopian tubes, the opposite way the sperm came to make its way to the uterus. As it does so, the same biomechanics that helped the sperm to the egg—the microtubules in the sperm's tail described in chapter two—are now involved in helping to guide the embryo back toward the uterus.

Along with flagella, which help sperm and other microorganisms swim, there are other hairlike structures that stick out of cells—cilia. They are generally smaller than the sperm's flagella, being around 0.25 micrometers in diameter and 6 micrometers in length—about the length of a sperm's head—and can be fixed to an immobile cell where they move the fluid that they are bathed in. Cilia are also found in several organs in the body. In the lungs, for example, cilia propel layers of mucus that line the airways. Cilia also line the fallopian tubes, where they are not uniformly spread as in the lungs but appear more like a forest with some clearings here and there.

Given their tiny size, cilia have the same tiny Reynolds number as sperm, and so they have the same issues when it comes to reciprocal motion to generate a unidirectional push on a fluid. To get around this, the cilia in the fallopian tubes have the same "9 + 2" microtubule structure as sperm, which allows them to make complex movements (see figure). They use this to move the fluid around the cell by performing what could be described as a one-handed breaststroke, with a power stroke different from the recovery stroke. The length of the cilia is well optimized for the job. Any shorter, and not

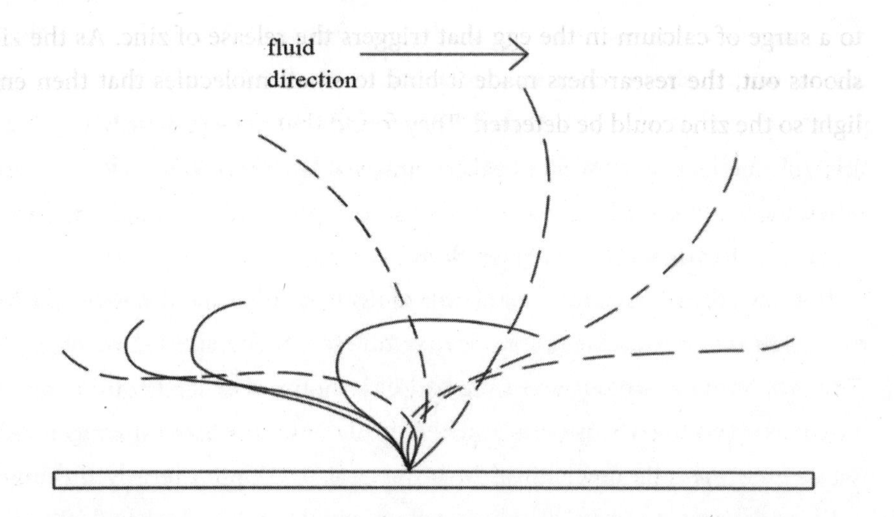

Fig 2 | *Cilia on the body of immobile cells can move the fluid they are bathed in by carrying out different "power" (outlined) and "recovery" (solid) strokes.*

enough flow would be created, while if too long they would be too "floppy" and, again, fail to create the necessary force on the fluid.

Despite their size, cilia manage to create enough of a flow in the fluid to move an egg cell or embryo. In 1982, the Australian mathematician John Blake,* who at the time was working at the University of Wollongong in Australia, showed that the fluid flow caused by the cilia is sufficiently strong to propel the egg in the fallopian tubes, even though it is widely thought that muscular contractions in the oviduct have at least the same, if not larger, effect.†

As the embryo is propelled toward the uterus in such a manner, a lot is changing. The cells in the embryo now begin to shift around and differentiate, so that by around day five it does not look like a smooth ball of cells

* Blake made several contributions to biological fluid mechanics, one of which was the mathematical description called a "blakelet"—similar to the Stokeslet—that describes the fluid flow around an object that is near a surface.

† Sometimes, an embryo doesn't make it to the uterus but still continues to develop and can embed in the fallopian tube instead. This is called an ectopic pregnancy and always needs to be terminated given the danger to the mother.

anymore. This "blastocyst" has two distinct features. One is the trophoblast that is located on the outer surface of the embryo and will help to form the placenta (see chapter eight) while the other is the inner-cell mass, which eventually becomes the baby. The remaining space is a fluid-filled compartment called the blastocoel.

To get to the blastocyst stage, however, requires the embryo to change from a smooth symmetrical ball of cells to the trophoblast and inner-cell mass region. In other words, the spherical symmetry of the embryo must break. After all, humans are not just a big ball of trillions of cells.* While scientists have been studying the genetic and chemical controls over symmetry breaking in embryos, the mechanical or physical aspects were largely unknown until work carried out in 2019 by developmental cell biophysicist Jean-Léon Maître at the Curie Institute in Paris and colleagues. By carrying out ultrafast imaging of mice embryos, they discovered for the first time the appearance of hundreds of tiny bubbles of water—each around a micrometer in size—between the cells. These bubbles were fleeting, which can explain why they had not been seen before. Remarkably, the researchers found that the bubbles could break apart proteins that held the cells together.

These individual bubbles are thought to originate from liquid outside the embryo, and as they flow between the cells, they collate to produce a single, large, water-filled cavity called the lumen.[9] As the fluid builds up in the lumen, it pushes the cells that become the fetus to one side to gather. Not only does this process play a critical role in helping the blastocyst to implant into the uterine wall, but it also begins the process of defining an orientation—the front and back of the embryo. The work by Maître's team was carried out in mice embryos, and they plan to do similar investigations on human embryos to see if the same mechanics are at play. Given that an inability for the embryo to embed in uterus is one of the biggest causes of pregnancy loss, the research may help assisted-reproduction clinics to identify which embryos have the best chance of implanting.

---

* For a fascinating description of how symmetry breaks on a cellular level, see Zernicka-Goetz, M., and Highfield, R. *The Dance of Life* (London: W H Allen, 2020).

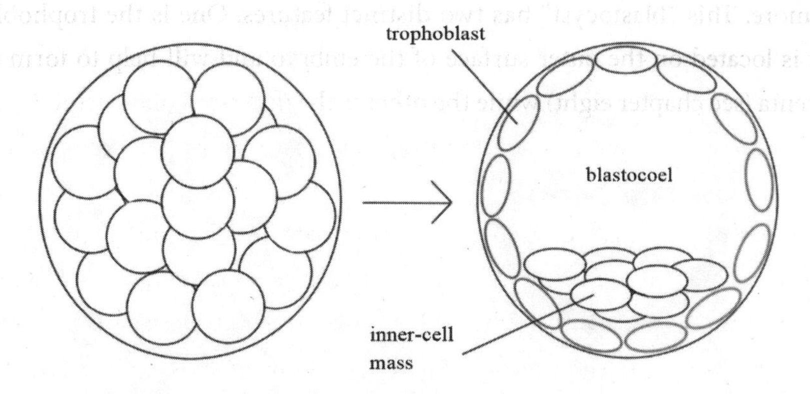

Fig 3 | *Following conception, the cells in the embryo continue to divide (left). Around day five it becomes a blastocyst, featuring the inner-cell mass and the trophoblast.*

Once this embryo differentiation occurs, at around day six, the blastocyst begins "hatching"—it literally bursts out of the zona pellucida*—and starts to implant itself in the endometrium† (the lining of the uterus). It then begins to lock into the maternal blood supply, much like a thirsty vampire (more on that in chapter eight). Once inside the endometrium, the embryo starts to form the axis and outline of the body. Snugly inside the uterus, the trophoblast remains on the outer region of the embryo, but the inner mass now starts its own migration and transformation, so by day twelve it forms two discs—dubbed the epiblast and hypoblast—that sit in the middle of the developing embryo a bit like the shape of an 8, with the bottom "o" the hypoblast and the top "o" the epiblast. In the middle, they meet, somewhat like the bread of a sandwich that, for the moment, contains no filling. The hypoblast will give rise to the yolk

* We don't fully know how this happens in the uterus given the difficulty of carrying out in vivo experiments; it may be that enzymes digest and dissolve away the zona pellucida.

† At that time in the menstrual cycle, the endometrium has become thicker from the secretion of progesterone by the remaining follicles in the ovary.

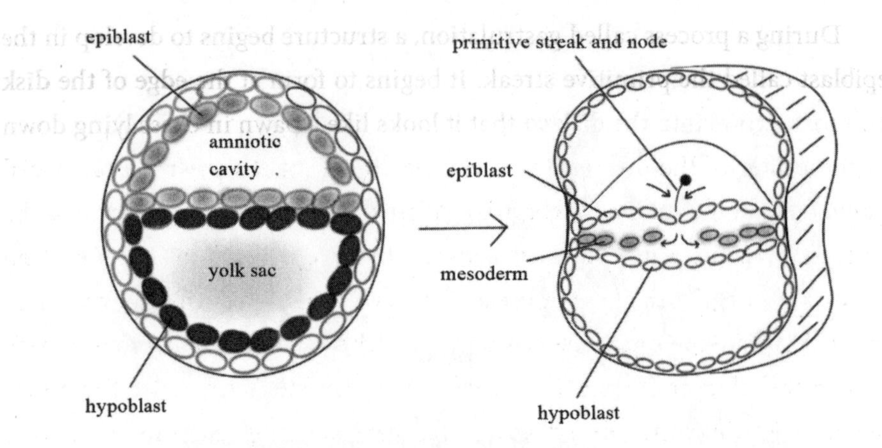

Fig 4 | *By day twelve (left), the embryo contains two layers, the epiblast and hypoblast. A few days later, the process of gastrulation helps the embryo to develop a body plan in which the primitive streak and node begin to form (right).*

sac, which initially nourishes the baby before the placenta takes over. On the other side is the beginnings of the amniotic sac that the baby will reside in.*

There are many incredible transformations that have already taken place, and more are about to happen before anyone even knows they are pregnant or has taken that pregnancy test. Around day fourteen, the two-layer sandwich-like disk of the epiblast and hypoblast begins to break symmetry so that the head and tail axis of the embryo starts to develop. This also marks the end point where experiments can be carried out on human embryos, mainly because the embryo is beginning to develop a central nervous system and represents the formation of a person—from a legal definition—rather than just a disk of cells.†

---

* For a much more thorough description of how a human embryo develops, see: Davies, J.A. *Life Unfolding: How the Human Body Creates Itself* (Oxford University Press, 2014).

† Some think that this fourteen-day limit, first proposed in 1979, should be extended to twenty-eight days so that some diseases, such as congenital heart disease, can be better studied as well as providing the hope of increasing the success of assisted reproduction techniques. In May 2021, the International Society for Stem Cell Research relaxed the limit, saying that studies proposing to grow human embryos beyond

During a process called gastrulation, a structure begins to develop in the epiblast called the primitive streak. It begins to form at the edge of the disk and then grows into the disk so that it looks like a pawn in chess lying down in the center of the disk (see figure). The "head" of the pawn has a special name—the node—and cells begin to gather there and then push through the epiblast to create a new layer in between the two disks, a bit like injecting some jam in between slices of bread. This eventually gives the embryo three distinct layers. The top is the ectoderm, which will form the nervous system and epithelial layer of skin. The newly formed middle layer, called the mesoderm, gives rise to the connective tissues, heart, and muscle tissue. Finally, the bottom layer, the endoderm, contributes to become the gastrointestinal and respiratory tracts.

In this whirlwind tour of embryogenesis, the embryo forms the axis of the body in the space of a couple of weeks. Having already established a front and back and head and tail, there is still one more major symmetry to break—the left-right symmetry. A human on the outside seems symmetrical in some sense. If you stand in front of the mirror and put an imaginary dividing line straight down the middle to cut you in half from head to foot—known as the sagittal plane—you would see relatively perfect symmetry.* Yet, as we know, on the inside asymmetry reigns, with the heart tilting to the left, the pancreas, stomach, and spleen lying toward the left while the liver is on the right. This handedness of organs is remarkably consistent, but there are instances where it is not.

Someone who first came across a case where things can be different was da Vinci. Despite being a great artist and scientist—but not quite getting the inner details of coitus correct—da Vinci had a few quirks. He created his own shorthand and even mirrored his writing, starting on the right side of the page and moving to the left. But even he must have been surprised one day

---

fourteen days could be considered on a case-by-case basis: https://www.isscr.org/policy/guidelines-for-stem-cell-research-and-clinical-translation.

* Although not quite, at least in men. A study in 1979 showed that in about two-thirds of men, the left testicle hangs lower than the right. The work, which analyzed the testicles of Greek sculptures, won the authors the 2002 Ig Nobel Prize for Medicine.

in the fifteenth century when he studied the torso of a dead woman. During dissection, he found that the heart was not tilted to the left, as it is for most people, but rather to the right. Just like his writing, it was a mirror image of a "normal" person's heart. This condition, now known as dextrocardia, affects less than 1 percent of the population, and those who have it are still able to lead a normal life. In most cases, they do not even know they have the condition.

While this must have been intriguing for an inquisitive mind like da Vinci, unknown to him, there is an even more extreme version than dextrocardia in which *all* the internal organs are mirrored. This was first discovered by the Scottish physician Matthew Baillie, who in 1788 published a detailed account of dissecting a cadaver that had transposed organs.[10] "Upon opening the cavity of the thorax and abdomen, the different situation of the viscera was so striking as immediately to excite the attention of the pupils who were engaged in dissecting it," Baillie wrote.

The condition was given the name *situs inversus*, from the Latin *situs* for "location" and *inversus* for "opposite." With an estimated occurrence rate of one in ten thousand births, some notable people who have the condition include the Spanish singer Enrique Iglesias and the US singer Donny Osmond. As with dextrocardia, it is perfectly possible to live a normal life and not even know that you have it. The question, then, is why—and how—does nature select the left so much for, say, the heart's orientation? After all, if it were random then you would expect half of people to have their heart tilted to the left and the other half having it tilted to the right. This problem turned out to be one the tools of fluid dynamics were well-suited for solving—and it was all thanks to those little cilia.

Casting your mind back to the developing embryo, you will remember that the node on the primitive streak happens to contain about two hundred to three hundred cilia that are about 5 micrometers in length. They are bathed in a thin film of embryonic fluid, which has a viscosity like saline and is present between all three layers in the developing embryo. However, these cilia have a slightly different internal structure than the cilia in the fallopian tubes and the flagella of the sperm that we have come across. In this case, they have what is called a "9 + 0" structure—the same nine microtubule structure in a ring, but without the central pair. This means that they cannot perform the same complex motion as a sperm's flagella, such as bending. Instead, they simply rotate from their fixed point on the cell, somewhat like putting one end of a stick in the ground and then moving the other end in a circle, which traces out an imaginary cone.

The importance of these cilia to breaking symmetry in the body was confirmed in experiments carried out on mice some twenty years ago. Scientists found that if the cilia on the node were disabled from moving, then some mice developed with their heart tilted to the left while some had it on the right—the outcome appeared random.[11] So, while it was clear that cilia are responsible for left-right symmetry breaking in the embryo, it was not known how this circular motion of the cilia could result in a leftward fluid flow.

A breakthrough came in 2004 from a mathematical analysis led by mathematician Julyan Cartwright, who was working at the Laboratory for Crystallographic Studies in Grenada, Spain. The cilia rotate with a frequency of

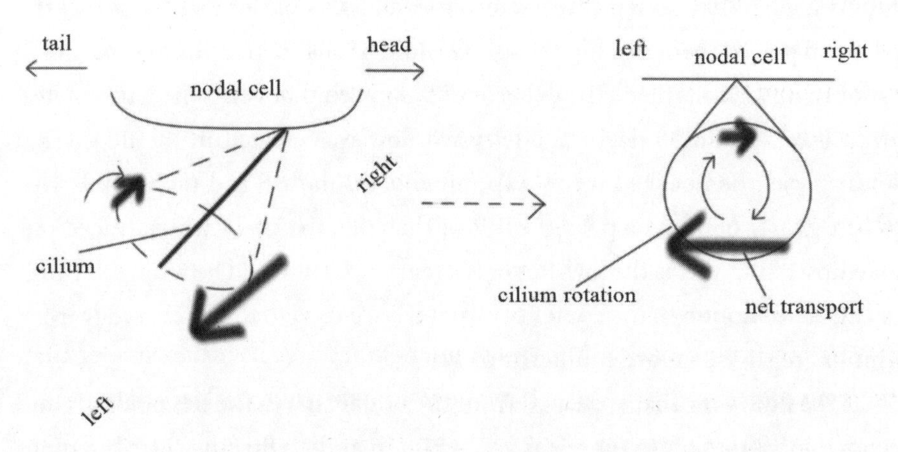

Fig 5 | *The cilia on the cells of the primitive node rotate in a specific way to create a greater left-field flow, which breaks the left-right symmetry in the body.*

about 10–20 hertz, similar to the rotational speed of an idling car engine. Cartwright and colleagues carried out simulations of the fluid dynamics that would be produced if the cilia were tilted about 35 degrees toward the already established posterior[12] and then rotated around this fixed point in a clockwise manner. The team found that in this setup, a left-biased flow could be created because the leftward motion results in a greater effect on the fluid than the rightward motions.

It's like looking directly at a propeller from behind a speedboat. As the cilia are pointed toward the back and rotate clockwise, they sweep the fluid to the left when moving toward the bottom of the imaginary circle. When they move toward the top of the circle, the cilia sweep the fluid to the right. The motion of the cilia toward the top of the circle is not as effective because the cilia are near the cell's surface (or the bottom of the boat), and so the surface of the cell resists the motion. At the bottom of the circle, however, the ends of the cilia are farthest away from the cell, so nothing stands in the way and the sweeping in this direction is more effective.

It is all well and good having a model, but as with the theory of Turing patterns, it had to be experimentally confirmed. This was done a year later when research groups in Japan mimicked this cilia orientation in a larger-scale experiment.[13] They used 6-millimeter-long (about a quarter inch) wires for the cilia and placed them in a highly viscous fluid to ensure that the low Reynolds number was maintained. The experiments showed that when the wire rotated at a tilt of around 30 degrees, a leftward flow was created in the fluid. This finding was then backed up by experiments on mouse and rabbit embryos, which clearly outlined a posterior tilt of the cilia, the tip of which traced out an ellipse.[14] But this is the low-Reynolds regime, so it would be wrong to think of this fluid motion in the context of what we are used to in our everyday lives. Things are slightly more counterintuitive.

"The fluid flow that is created from the nodal cilia in the Reynolds regime is not a constant flow to the left as you might imagine a flowing river, but more like two steps forward and then one step back," mathematician David Smith at the University of Birmingham explained to me. Smith, who studies cilia in the lungs and on the embryonic node,* earned his PhD in mathematical biology at the same university in 2005. In this sense, the fluid moves to the left, followed by a bit to the right, and then more to the left—leaving the net effect of "stuff" being deposited in a greater volume to the left.

It is thought that the floor of the cells on the node produce proteins, one of which is called "nodal" after its origin. This is a powerful signaling molecule that affects gene expression to switch on certain aspects of development.[15] The proteins are moving all around the developing embryo, switching on genes to form the components of the human body, such as the heart, spine, liver, and lungs. There is a lot we still do not know about left-right symmetry breaking beyond a basic description of how it happens, but it is now clear that those little nodal cilia play a critical role in the specific orientation of organs, such as the heart and liver. Just like Turing's example of morphogen patterning some fifty years earlier, how left-right symmetry is broken in the embryo is another perfect demonstration of the power of mathematics to correctly predict behavior in biological systems, despite their complexity.

Over the past decade, Otger Campàs, a biological physicist at the University of California, Santa Barbara, has been studying some of the forces involved with how the embryonic body lengthens—the next stage after all the structures are in place. Measuring forces in cells or tissue has often involved doing so outside of the body in a petri dish, which can provide some information. But to get precise measurements in a living animal developing so rapidly required Campàs to come up with a new technique. Rather than study mouse embryos, however, he examined zebra fish, which happen to be an excellent research model for several reasons. One is that they are vertebrates. They are also

* Smith's work focuses on the left-right organizer in the zebra fish, which is called Kupffer's vesicle. Given the more complex arrangement of cilia in the vesicle, the fluid mechanics is more difficult to study and is still an active area of research.

abundant and cheap, but more importantly, they are almost completely transparent, which allows us to clearly see the development of internal structures.

When Campàs was a postdoctoral researcher at Harvard University in the late 2000s, he came up with the idea of measuring the forces by injecting drops of oil between the cells in the zebra fish's tail, which can double in length in under five hours. After carrying out some preliminary investigations—which involved simple supermarket-bought olive oil—Campàs moved to Santa Barbara to build up his own team. He then began to develop the technique further, this time using biocompatible oils, coating the droplets and learning how to load magnetic nanoparticles into them, all of which took around eight years to achieve.

To advance the research, the team injected single oil droplets, each loaded with a magnetic nanoparticle, into the space between the cells in the tail of the zebra fish. Then they applied a magnetic field to deform the droplet. This, in turn, deformed the tissue. The researchers investigated whether the droplets could return to their spherical shape and what amount of stress was needed to permanently deform the tissue. All this information enabled them to generate a picture of the mechanical properties of the fish's cells and tissues. By analyzing how squashed the particles became, they could get a measure of the stress applied to the droplet, and by placing droplets in different places, they could then map out the stresses in the tail to measure how densely packed the cells were.

Campàs and colleagues found that the cells at the end of the growing tail are like a fluid, in that they can flow freely past each other, and the tissue could be deformed more easily. They suggest that cells travel along the surface of the tail toward the tip before "reentering" the tail to bring more material to elongate it. When moving from the tail toward the head, however, the cells became more fixed and rigid, and it is here that cells began to form structures that eventually give rise to the animal's vertebrae.[16] As the cells crowd together, they get locked into place like a solid—somewhat like cars on a highway suddenly approaching a traffic jam and getting tightly packed. This is known as a "jamming" transition and pops up in other biological systems, such as in wound healing and when cancerous tissues form.

The team also added a fluorescent dye between the cells in the tail and found that cells at the tail's tip have much more space than at the other end, with the cells "jiggling" more at the tip than toward the head. This combination of jamming and a lack of jiggling toward the head conspires to solidify the tail, with Campàs comparing it to glass blowing, liquifying a part that needs sculpting and then letting it solidify. "This same process is exactly how D'Arcy Thompson described morphogenesis, although without any evidence at the time," Campàs said. "Yet, this is what we found in the formation of the zebra fish's tail." Although the work has only been done in zebra fish embryos, Campàs, who is now based at the Technical University in Dresden, Germany, expects liquid-to-solid transitions to be a general feature of morphogenesis for how vertebrates form, including humans. "Within the next twenty years, we will learn a lot about the physics of living things," Campàs said. "It's an exciting time."

In the space of a few weeks, a human embryo goes from a clump of cells at day six to an initial outline of the body such that by the time you know you are pregnant—probably through a pregnancy test, which we will look at next—the basics of a body plan are already taking shape. That rapid development continues, so that by week six, it is possible to see the early signs of a heartbeat on a vaginal ultrasound scan, and by the end of week twelve of gestation, the main outline of the body is pretty much fully formed.

Getting a better understanding of these processes, whether in humans or other animal models, requires collaborations between scientists from different fields of study. But the importance of physical science in helping to elucidate these mysteries is now no longer in question and represents a powerful tool in studying the shaping of living things. As Thompson wrote over one hundred years ago in *On Growth and Form*, "Of the construction and growth and working of the [human] body, as of all that is of the earth earthy, physical science is, in my humble opinion, our only teacher and guide."

PREGNANCY PROBABILITIES AND TESTS

How easy is it to get pregnant? After all, anxious parents regularly reinforce to their teenagers, "It only takes one time!" At some point in a young adult's life, the focus might be on trying *not* to get pregnant via contraception or perhaps abstinence. When you finally reach the stage where you want to start a family and bring another human into the world, you may find that you and your partner spend months trying with nothing to show for all your efforts. But weren't we told in sex-education classes and constantly by our parents that it is easy to get pregnant? Is it all a scam by the contraception industry?

It is a seemingly simple enough question. What is the chance in a given month of becoming pregnant? It is a tricky question to answer on a person-by-person basis. The best attempt so far comes from mining the detailed records provided by couples as they tried to conceive. One such program conducted in the early 1980s involved 221 couples in North Carolina who noted when they had sex and when the woman ovulated, marked by a surge in luteinizing hormone corresponding to the release of an egg from a follicle. In another, conducted in Italy between 1993 and 1997, 193 women documented when they had sex, together with observations of their cervical mucus, scoring it in one of four numbered categories: dry (one); humid or damp feeling (two); thick, creamy, white mucus (three); or slippery, stretchy, watery clear mucus (four).

The biggest study in the 1990s—the European Study of Daily Fecundability—followed 782 European women as they attempted to get pregnant. Researchers collected data on over 7,200 menstrual cycles, in which women kept records of when they had sex as well as their so-called basal (resting) body temperature and a description of their cervical mucus. In the case of following a woman's basal body temperature, it rises just under half a degree Celsius at ovulation, which can be detected with a sensitive enough thermometer. All studies captured the key parameter: whether a pregnancy was achieved after all this effort.

These surveys provided a huge amount of data, but scientists needed a way to dig out any trends. The issue is that in a conception cycle there can be several days with intercourse, so it is difficult to know which one results in conception. However, by using simple rules of probabilities, it is possible to attribute some likelihoods to each day. The first attempt at applying these techniques was by researchers from the United States who in 1969 used data collected by 241 British married couples in which the women recorded their basal body temperature and noted down when they had sex.[1] Using statistical techniques to determine the probabilities during the cycle that would result in a successful pregnancy, they found the ideal time for intercourse was two days before ovulation, with a probability of conception of 30 percent, while after ovulation the probability fell to almost zero. This "likelihood" model was refined in other studies, including one in 1980 that used the same data and found the probability of pregnancy for at least a six-week duration was 49 percent when having intercourse every day.[2]

In the 1990s and 2000s, statistician David Dunson and his colleagues at Duke University took a slightly different approach, using a powerful technique called Bayesian statistics* to analyze the female menstrual cycle and the

* Formulated in the 1700s by the English statistician, philosopher, and Presbyterian minister Reverend Thomas Bayes, Bayesian analysis is a technique that calculates the probability of an event happening by considering not only the likelihood of it happening

best times to conceive. It is well known that cycle lengths can vary between women and even for the same person from month to month. But the statistical analysis showed that even for women with a regular cycle of twenty-eight days, the point of ovulation could still shift. By examining around seven hundred cycles that were twenty-eight days long in the North Carolina data, only 10 percent had ovulation at day fourteen—the day you would normally expect ovulation to occur.[3] They also found that the fertile period in the cycle was six days in length—five days before ovulation and the day of ovulation itself.

As earlier studies had shown, the probability of conception a day after ovulation was practically zero, so not much point trying to get pregnant at that time. The study also confirmed that timing is key, with the chance of getting pregnant as much as 30 percent just one day before ovulation occurred. Another interesting take from the North Carolina study was that the frequency of sex in the couples peaked at ovulation. Taking the mean daily frequency of intercourse among the participants in the study, Dunson and colleagues found that during the six fertile days, the frequency of intercourse was 24 percent higher than on all other nonbleeding days. The amount of sex then plummeted just one or two days following ovulation.[4]

A Bayesian analysis of the Italian study, led by Bruno Scarpa at the University of Padua in Italy, found that if couples disregarded the indicators from cervical mucus as well as the menstrual calendar and just had sex once a week, then the average waiting time for a pregnancy was four cycles.[5] However, if they observed their mucus but *only* had sex when it had a score of four and between days thirteen to seventeen, then the average number of required intercourse days was 2.42, with a conception probability of 35 percent. This means that it would take, on average, three cycles to become pregnant. For couples who were less picky about the mucus score and had sex when it was

---

but also information about what previously occurred, what is known as "prior beliefs." These two aspects are combined to produce a "posterior belief," and this is iterated based on new evidence until a good estimate is found. Bayesian statistics is a particularly powerful method to pick out trends when data are noisy, with the technique used extensively in medicine and genetics.

three or four between day thirteen to seventeen, the probability of conception was 47 percent, which means waiting for only two cycles. So, for those paying attention to the calendar, having sex every other day can be sufficient. But what the Bayesian analysis shows is that if you really want to optimize how many times you need to get it on, then paying attention to cervical mucus signs and the calendar is a powerful way to achieve pregnancy quickly.

The fertility time bomb is something you hear a lot about in the press as well as from your eager-for-grandchildren parents. But with many more people holding off from having a child until their late thirties or early forties, is there really an issue? After all, you still see older parents pushing buggies around—just make sure they are the parents before making comments about how nice it is for grandparents to spend quality time with their grandchildren.

Using data from the European study, Dunson and colleagues found that women aged nineteen to twenty-six had a significantly higher probability of becoming pregnant, as expected, than those aged twenty-seven to thirty-four, while women aged thirty-five to forty saw a further decline.[6] The percentage of women failing to conceive within twelve cycles (if they have sex around two days each week) was 8 percent for nineteen- to twenty-six-year-olds, 13 percent for those aged twenty-seven to thirty-four, and 18 percent for those aged thirty-five to thirty-nine. So, while the probability does drop, it is hardly a cliff edge by age forty compared to the early twenties.

The study also looked at the impact of male age, finding that while there is no impact for men under thirty-five regarding fertility, above it there is. If both the male and female are thirty-five, then, as for females mentioned above, the probability over twelve cycles of failing to become pregnant is 18 percent, but if the man is forty or older, then it becomes 28 percent—a big jump for just a five-year gap. The proportion of couples who were sterile was 1 percent, with no impact on age, revealing that the age-related drop in fertility is due to a gradual decline. The work shows it is unlikely that a couple is sterile and so cannot conceive naturally at some point. So, while doctors might say keep trying for twelve months before thinking about fertility treatment, it is really borne out in the Bayesian analysis.

The powerful statistical techniques of Bayesian analysis are now finding applications in pregnancy apps, which claim to be able to accurately spot ovulation with just a daily input of basal body temperature. This knowledge can then be used to either have sex on or before that day or abstain as a form of contraception. These apps, in turn, are also generating huge amounts of data that can be studied. Work in 2019 by the people behind the Natural Cycles app analyzed over 600,000 cycles from almost 125,000 users .[7] They found that the mean cycle length was 29.3 days and that the mean cycle length decreased by 0.19 days each year of age, from twenty-five to forty-five years.[8] The mean variation of cycle length was 0.4 days for women with a body mass index, or BMI, greater than 35—14 percent higher than those classed with a "normal" index of 18.5 to 25. It is likely more analysis will emerge from these apps to further shine a light not only on the menstrual cycle but also on fecundability, or the probability of becoming pregnant within one menstrual cycle.

Once you and your partner have tried to get pregnant, the first port of call will be to see if it has been successful. When my wife, Claire, and I found out we were expecting our first child, we were over the moon. I still remember staring at the faint blue line on the pregnancy test one early morning in disbelief. The more I looked at the mark, the darker it became, although it was still very much lighter than the neighboring "control line." The test was done a few days before the anticipated period, so we questioned if our minds were playing tricks on us. Given the "two-for-one" discounts on pregnancy tests in supermarkets, my wife did a few follow-up tests after that first one. It was a bit of harmless fun, although slightly expensive, to see the line becoming more pronounced. After all, it made us come to terms with the fact that the pregnancy was happening so, all going well, we could plan what box sets to watch at 3 AM in nine months' time.

That very first test was done in early December, and over the holiday period, we were staying with family when my wife decided to do another one. After all, a nice blue line would make a great Christmas present. The procedure was correctly performed with the first-morning pee on Christmas Day and the control line began to turn blue. We waited for the other line to emerge. And then waited . . . and waited. Nothing came. After a while, a very faded line appeared just as it did when we carried out that first test a month or so before but did not continue to darken. We both looked at each other and did not say a word. Inside, our minds were racing. After all those strong lines, could it be that we were losing or had lost the pregnancy? We were devastated. What a dreadful start to Christmas.

But it would turn out fine in the end—all thanks to a particular peculiarity with modern pregnancy tests.

Pregnancy tests today are simple to use and can give results in minutes. But that was not always the case. The first recorded use of a test for pregnancy is thought to have been carried out in around 1350 BC by the ancient Egyptians. We know this because archaeologists discovered documentation, written in hieroglyphs, stating the rules for the test. Over several days, women urinated on various grains, such as barley and wheat. If a seed germinated and grew, then it meant the woman was pregnant and the method could, it was claimed, even tell the sex of the child. If the barley sprouted, then it was a boy, and if the wheat did, it was girl.

Research carried out in the 1960s to reproduce the method with the urine of pregnant women showed that the seed test had an accuracy of about 70 percent for predicting a pregnancy—not bad for such a primitive method (although its ability to tell whether the baby would be a boy or a girl could not be replicated, so do not get any ideas for your next gender-reveal party).[9] The test worked because elevated levels of estrogen in the woman's urine stimulated the grains to sprout.

It took hundreds of years of women peeing on wheat and barley for a more accurate test to come along. This time, it relied on picking up another key pregnancy hormone, human chorionic gonadotropin (hCG). We now know that this is a much better hormone to use, as it is produced during pregnancy and made by trophoblast cells on the outer surface of the blastocyst that go on to make the placenta.* Following the embryo's implantation into the endometrium of the uterus, hCG initially doubles every couple of days, peaking roughly ten weeks following conception.

The hormone was first discovered in 1928 by the German gynecologists Selmar Aschheim and Bernhard Zondek in the urine of pregnant women. The duo found it had a powerful response in other animals. When they injected 3 milliliters (about a tenth of an ounce) of urine twice daily for three days into five female mice, they found it could stimulate the rodents' ovaries, causing them to go into heat. Enlarged ovaries indicated pregnancy, but if the ovaries remained their normal size, then pregnancy likely did not occur. Unfortunately, to detect if there had been any response, the mice had to be killed and the ovaries dissected and examined. The test, named A-Z after the researchers' initials, was a big improvement to sprouting seeds—correct in about 98 percent of cases.[10]

Around the same time, American medical doctor and reproductive physiologist Maurice Friedman, at the University of Pennsylvania Medical School, found a similar response in rabbits. Two rabbits were usually used for the test and were injected with about 8 milliliters (about a quarter of an ounce) of human hCG into their ear veins. The presence of hCG caused certain formations, known as corpora lutea and corpora hemorrhagica, in the tissues of the rabbits' ovaries. Corpora lutea is a small yellow mass that releases hormones such as progesterone and forms near the follicle in the ovary that has released the egg. Corpora hemorrhagica, meanwhile, develops after ovulation when the follicle collapses and is filled with blood that then clots.

---

* A slight caveat is that hCG does not always mean a successful or viable pregnancy. An ectopic pregnancy, in which the embryo implants into the fallopian tubes, would result in a "false positive" as would the tumors of some forms of cancers, such as testicular cancer, which can produce hCG. In principle, a pregnancy test could tell a man with testicular cancer that he is "pregnant."

Unfortunately, as with the A-Z test, to detect if this change had happened, the rabbits had to be killed and their ovaries examined. But the method was quicker than the mice test, as it only took a day to carry out.* "It's highly reliable," Friedman told the *New York Times*.[11] "The only more reliable test is to wait nine months."

The butchering of scores of rabbits and mice felt somewhat unnecessary, and a more humane approach was sought. Enter the English zoologist Lancelot Hogben. He was working at the University of Cape Town in the late 1920s studying the African clawed frog (*Xenopus laevis*) with a focus on the animal's production of hormones. When Hogben returned to the United Kingdom in 1930, he brought back a colony of these frogs and sent some to his colleague Francis Crew at the University of Edinburgh in Scotland. In a series of experiments, Crew injected the urine of pregnant women into the frog's dorsal lymph sac (located on the top of its body), finding that after around twelve hours, the frog spawned eggs or produced sperm. This method was incredibly accurate to determine pregnancy, with a success rate of 99.8 percent.[12] It also gave an answer much faster than previous methods and crucially did not involve killing the frogs in the process, allowing them to be reused. "Thank you for your report on the pregnancy test on Mrs. [X]. You may be interested to know that of one general practicioner (GP) of many years' standing, one specialist gynecologist and one frog, only the frog was correct," a doctor wrote to Hogben to praise the new method.[13]

Yet there was a dark side to the frog test. Given its speed and reusability, tens of thousands of frogs were tested over decades, and when they became surplus, they were simply abandoned in the wild to form their own colonies. The bigger problem was that they brought a deadly fungus with them—*Batrachochytrium dendrobatidis*—that affected native frog species, which had never been exposed to the pathogen before. It is estimated up to two hundred species of frogs worldwide were wiped out because of the fungus invasion.[14]

Time was up on the use of animals for pregnancy testing.

---

* This is the origin for the euphemism "the rabbit died," to indicate that someone was pregnant.

Those who have young children will know how efficient they are at attracting a plethora of bugs that they then kindly bring back home to infect everyone else. In the winter months, kids are constantly sneezing or just generally oozing mucus as their developing immune system fights off illness. When an infection happens, the body's immune system encounters the invading virus and sets out to expel it. One method of defense is to produce antibodies, small fork-like proteins that bind to the virus's receptors to inhibit its potency. Sometimes, these proteins hang around in the blood system for a while after the virus is gone, so if it is encountered a second time soon after, then the antibodies will swiftly recognize it and raise the alarm, giving the immune system the upper hand, which is how some vaccines generally work.

In the 1960s, scientists began to develop pregnancy tests to detect hCG in urine, using antibodies that could bind with, and therefore detect, hCG molecules. Tests that detect the presence of a certain molecule via antibodies are known as "immunoassays," which were first introduced in 1959 to detect insulin, and since then have been used to detect cancers, illegal drugs, viruses, and other microbes.* Using an immunoassay as a pregnancy test required a bit more than just adding urine to a solution containing hCG antibodies. The tests in the early 1960s featured the red blood cells of rabbits that were coated in the lab with hCG antibodies (so, not yet totally free from animal products). The other main component of the test was red blood cells from sheep coated with hCG. These two components were put together, and if the woman's urine did not contain hCG, then the antibodies from the rabbit blood would bind with the hCG on the sheep blood to result in a clump of blood. If the woman's urine contained hCG, however, then it would bind with the rabbit antibodies, causing the sheep blood cells to fall out of the solution. The sheep blood cells would then form a red-brown ring on the bottom of the container. As a sprout

* The immunoassays industry is big business, worth an estimated $18 billion in 2020 and estimated to grow to $25 billion by 2025, according to Grand View Research: https://www.grandviewresearch.com/industry-analysis/immunoassay-market.

of barley was a sign of pregnancy in ancient Egypt, in the mid-twentieth century, it was a ring of sheep's blood.

Given that any movement of the container would spoil the result, the tests were initially carried out in a carefully controlled lab setting. Technicians would line up the containers with the blood-urine mix and place them directly above a mirror so they could easily see if the rings were present. This test took about two hours to complete, and if done around twenty-two days after the estimated day of ovulation, it was about 99.8 percent accurate.[15] The only downside was that it involved going to your doctor and giving a urine sample and then waiting a week or two for the results to come back.

A testing revolution in the 1960s began thanks to Margaret Crane, who was hired by Organon Pharmaceuticals in New Jersey as a freelance graphic designer to work on the packaging design for a new line of cosmetics. As she toured the firm's pregnancy-testing labs, she encountered the rows of vials suspended above mirrors. Upon learning that they were pregnancy tests, Crane was convinced that the process could be miniaturized and standardized enough to be done at home. She began designing a home pregnancy test that contained all the basic elements of the laboratory test. She pitched it to Organon executives, who eventually took on the idea, and the firm was awarded a patent in 1971, listing Crane as the inventor.* Despite some resistance from doctors at the time about doing such tests at home, the first home-testing kit became available in US pharmacies in 1978, costing around $10.† Research by Organon showed that the home version was about 97 percent accurate—not quite as good as the lab version, but at least it could be done in the privacy of your own bathroom.

In the following years, scientists began working on tests that did not involve animal blood by finessing the design of synthetic antibodies that could pick up hCG. What makes modern pregnancy tests so accurate is that

* Crane assigned the rights of the patent to Organon for $1, so she didn't become rich because of it.

† In 2015, the Smithsonian Institution bought the original prototype, along with the first consumer version of the test, for $11,875 at a Bonham's auction.

they take advantage of two particular areas, called epitopes, on the hCG molecule that an antibody can attach to. These two hCG subunits—one called alpha-hCG and the other beta-hCG—were discovered in 1970 by researchers at State University of New York at Buffalo.[16] The alpha subunit is identical to the alpha subunit of other hormones, such as the luteinizing hormone that's produced in the anterior pituitary gland in the brain and is an important hormone for reproductive function in both men and women. The beta subunit, however, is unique to hCG.

Technological advancements eventually allowed placing the components of the test on a simple strip material. In 1988, Unilever introduced its "one-step" hCG sandwich immunoassay. One end of the strip absorbs urine (either by being dipped in urine or literally urinated on), which first filters it to make sure only the urine passes through. The urine soaks through to the opposite end of the strip, arriving at the so-called "reaction zone." This part features mobile antibodies that connect to the alpha subunit of hCG. These antibodies also have latex beads attached to them, which are dyed blue. The hCG-antibody combo then travels with the urine to the next area—the test zone*—where all the action happens. Secured to this zone (so not being able to move) are more antibodies, but these bind with the beta subunit of hCG. As this antibody is immobile, the hCG is stopped in its tracks. As more and more antibodies get trapped, the beads that they carry begin to accumulate and it is this that produces the blue line. The hCG molecule is sandwiched between two antibodies, and this explains the name "sandwich" immunoassays.

If there is no hCG present in the urine, the antibodies with beads simply travel on and do not connect with the stationary antibodies, resulting in no blue line. The antibodies now head to the control zone that also contains immobile antibodies. These, however, are designed to stick to the mobile antibodies, not the hCG. They connect with the antibodies, and then the beads again accumulate to produce a blue control line. So, regardless of whether the mobile antibodies pick up hCG or not, the control line should always be

* It is slightly more complicated than this picture paints. These are also "blocking molecules" on the strip that stop imperfect binding.

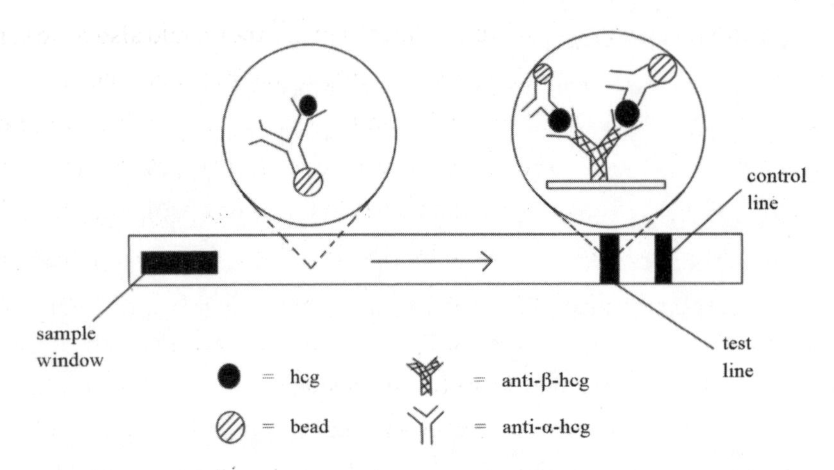

Fig 1 | *Schematic of a modern pregnancy test showing what happens when hCG is present in the urine. Adapted from [17].*

at least a pale blue. If it is not, then the test is faulty. Sandwich immunoassays are the quickest and easiest way to test, and hCG remains the best method we have of detecting early pregnancy.* As one TV commercial in 2006 for Clearblue Easy put it: It's "the most sophisticated advanced piece of technology you will ever pee on."†

Companies claim that their modern-day pregnancy tests can predict pregnancy up to six days before the expected missed period. But, despite these claims, there is always a limit. Levels of hCG at four days before the period is due are, on average, approximately 0.5 micrograms per liter—too low to produce a reliable signal, resulting in an accuracy of 55 percent even in modern pregnancy tests. That chance improves to 86 percent just a day later, becoming 99 percent accurate on the day of the period, when levels of hCG are, on average, around 10 micrograms per liter.[17] At four days following a missed period, the levels of hCG are 80 micrograms per liter. So, it pays to wait a little no matter how desperate you are to see the results.

---

* Of course, blood tests can also be used to detect hCG and give an absolute value, which pregnancy tests can't produce.

† Although it was later changed to "the most advanced piece of technology you will ever . . . well, you know."

And, too much enthusiasm to test later in pregnancy could also have some rather unintended consequences, which brings us back to that pregnancy test on Christmas Day 2014. Before Claire and I joined the festivities we did a quick consultation via "Dr. Google." It was then that we learned about the "hook effect" that can affect immunoassays. In terms of a pregnancy test, this is when there is such a high level of hCG in the blood that it overpowers the test. In this case, there are so many hCG-bound antibodies that they crowd the immobile antibodies in the test line and fail to attach properly. It's a bit like a mass of people trying to run toward a single door that just results in a deadlock. As many of the hCG antibodies fail to properly "sandwich" with the immobile antibodies, the beads fail to be delivered properly and so the line becomes fainter than it perhaps was before. This need to form a "proper sandwich" is not as necessary for the control line's antibodies, and so this part of the test still works fine.

From that moment on, we promised no more tests and to instead try to enjoy Christmas. After all, it would only be a few more weeks before we would have the ultimate proof that we had a baby on the way—the ultrasound examination.

FIRST INTERLUDE

THE PHYSICS OF ULTRASOUND SCANS

Going for a routine ultrasound scan during pregnancy is a nerve-racking experience. I remember it vividly the first time we did so. Claire and I were sitting in the hospital waiting room with several other couples who were all eagerly anticipating what they were about to see. There was a nervous apprehension in the air. After what seemed like an eternity, a nurse walked into the waiting area clutching a folder. "Banks," she shouted as we both stood up rather too eagerly to follow her.

That was just for a blood test, so we had to come back out again. But once called a second time, we were finally led to the ultrasound room. Following some pleasantries, Claire hopped onto a stiff bed, lay down, and lifted her top to reveal the abdomen. "This is going to feel a little cold," the sonographer warned, as she squeezed a thick, colorless gel onto her stomach.

The nurse placed a small handheld probe onto the gel and pressed it firmly onto the skin while moving it slightly from side to side. Our eyes turned to the monitor, transfixed. It was the first time we "saw" our son-to-be as he wriggled on-screen, heart beating away. The nurse slid the probe around, taking the odd screenshot to do a few measurements. "All looks lovely," she said as she offered her congratulations and printed off a few of the shots so we could take them home and stick them on the fridge. We were barely in the room ten minutes, but I could have stayed for hours. The nurse carried out the same procedure that day for a conveyor belt of expectant parents, all hoping for the same good news.

As one, quite rightly, marvels at that little life moving around its cozy home in the uterus, it is easy to have missed something equally amazing during the scan. While techniques such as MRI require huge magnets and are expensive to operate, ultrasound can be done relatively cheaply, safely, and quickly, while still able to produce incredibly detailed images of a fetus in situ. Rather than pulling protons in the body, as with MRI, ultrasound uses sound waves. But these are not ones that we can hear, having a much higher frequency, or shorter wavelength.*

There is no fundamental difference between sounds that we can hear and ultrasound; it is just that its frequency range is higher. Your ear detects sound when moving air particles cause the eardrum to vibrate. Humans can hear a wide range of frequencies—from around 30 to 20,000 hertz, but we are best at hearing sounds between 1,000 and 5,000 hertz.† Some animals, however, can hear at much lower frequencies than humans—infrasound—having eardrums that vibrate at those frequencies. Elephants, for example, can hear sounds at around 14 hertz while whales can hear at 7 hertz. Other animals can hear at much higher frequencies than we can.

Dolphins can hear up to 160,000 hertz while the greater wax moth can sense sound frequencies up to 300,000 hertz. The real ultrasound specialists, however, are bats, which produce it via their voice boxes or clicks of their tongues. Given that ultrasound is much better at reflecting off objects than the sound range we can hear, bats use it to determine the size, location, texture, and even speed of moving prey. All this information processing happens in a fraction of a second—and for a multitude of objects simultaneously—to allow

* Wave frequency is the number of waves that pass a fixed point in a given amount of time.

† Hertz is the unit of frequency, which is the number of cycles per second.

bats to obtain a three-dimensional picture of the world around them that is continuously updated in real time. So sophisticated is this "echolocation" that in some sense it is more effective than sight.* While we often say to someone that "You're as blind as a bat," a bat would say in turn (if they could speak), "You're as deaf as a human."

The first artificial production of ultrasound in the lab came in 1880 by the French physicist Pierre Curie† and his brother Jacques Curie. They discovered that certain materials, such as quartz and topaz, could produce a voltage when squeezed, which became known as the piezoelectric effect. The following year, the French physicist Gabriel Lippmann‡ proposed that the reverse should hold true as well: applying an electric field to a piezoelectric crystal could deform it, and this was soon demonstrated by the Curie brothers.

This inverse piezoelectric effect means that when applying an alternating current to a material, the crystal undergoes a repeated compression and relaxation, vibrating the air rapidly to produce ultrasound. The power of piezoelectric crystals is that the same crystal can act as both a transmitter and detector,

---

* In the late eighteenth century, Lazzaro Spallanzani, the bishop of Padua in Italy, conducted several experiments on different mammals as they navigated the dark. He noticed that when he put owls in a pitch-black room, they refused to fly. When he brought bats in, however, they flew just the same amount as if it were light. In the dark, they could even dodge elaborate patterns of wires that had bells attached to raise the alarm if they hit them. Investigating further (look away now, those concerned about animal welfare), Spallanzani blinded the bats with a red-hot needle, but amazingly they still managed to dodge the wires and bells. Spallanzani could only get the bats to hit the wires if he blocked their ear canals with brass tubes that had closed ends. Once he opened the tubes, the bats again regained their wire-dodging capability. How bats navigated remained a puzzle for another 150 years and was eventually solved by a Harvard University biology student named Donald Griffin, who used an ultrasound microphone to pick up signals emitted from flying bats, finding that the bats emitted ultrasound from their mouths but detected it with their ears.

† Pierre Curie is most famous for his work in radioactivity, for which he shared the 1903 Nobel Prize in Physics with his wife, Marie Curie, and Henri Becquerel, who first discovered the effect in 1896.

‡ Lippmann was also a physics Nobel laureate, awarded the 1908 physics prize for inventing a photographic technique to record and reproduce colors.

with the inverse piezoelectric effect used to produce ultrasound from an electrical signal while the piezoelectric effect is used to transform the ultrasound back into an electrical signal.

The first application of piezoelectric technology came via one of Pierre Curie's former students, French physicist Paul Langevin, who was a pioneer in crystallography and magnetism.* On the back of the loss of the *Titanic* in 1912, scientific efforts turned to ways to detect underwater structures, and with funding from the French government, Langevin built the world's first "hydrophone" in 1915. This featured a transducer that had a mosaic of thin quartz crystals glued between two steel plates.† By emitting high-frequency pulses (having a resonant frequency of about 150,000 hertz) from the transducer and measuring the amount of time it takes to hear an echo from the sound waves bouncing off an object, it was possible to calculate the distance to that object. What made this even more effective was that it could be housed in a submarine and used to locate the distance of enemy craft. Despite the promise of piezoelectric materials in the late nineteenth century, it took over half a century for ultrasound technology to be widely used in medicine. But once it took hold in the medical field, it quickly found applications in cardiology, urology, and cancer detection.

Then it moved into obstetrics, the field of medicine concerned with pregnancy and childbirth. Spotting a fetus during an abdominal scan is somewhat like detecting and locating a submarine underwater. When the ultrasound waves travel into the body, they pass through until they reach a boundary, say between soft tissue and bone. Some of the sound waves are reflected while others travel on farther until they reach another boundary. The reflected waves are picked up by the piezoelectric transducer located in the probe and relayed to a computer that calculates the distance from the probe to the tissue

---

* Langevin reportedly had a relationship with Marie Curie in 1910, around four years after her husband, Pierre, died after being hit by a wagon while crossing the street.

† Transducers are devices that convert one form of energy into another. In this case, the mechanical energy is converted into electrical energy.

or organ (boundaries) using the speed of sound in tissue (around 1,500 meters per second) as well as the time of each echo's return (usually on the order of millionths of a second). The ultrasound machine then displays the distances and intensities (related to the amplitude) of the returned waves on the screen, forming an image.

The world's first workable ultrasound scanner for obstetric use was created in 1958 by Ian Donald, who was a professor of midwifery at the University of Glasgow.[1] In the setup, the patient's abdomen was smeared in olive oil to make sure there was no air between the transducer and the skin that could affect the signal's quality. Then the probe, which consisted of two piezoelectric crystals made from barium titanate—one for receiving and one for transmitting—was moved from one side of the stomach to the other, rocked slightly from side to side by the technician during the scan.

One of the immediate benefits of ultrasound imaging was determining the location of the placenta (the physics of which we will cover in chapter eight). Placenta previa, a condition in which the placenta lies fully or partly over the cervix, was a significant cause of maternal death due to severe bleeding in late gestation. The best technique pre-1950s to determine where the placenta lay in the uterus involved intravenously injecting radioisotopes, such as radioactive sodium, into the mother's bloodstream. As the radioisotopes moved around the body, they emitted radioactive particles that could be detected with a scintillation counter (a device for detecting radiation through recording faint light pulses). When the radioisotopes made their way through the placenta, physicians could measure the radioactivity and crudely draw an outline of the placenta. The technique was not without shortcomings, given that it could not accurately detect the edge of the placenta, which meant that a scan might show the location of the placenta to be free from the cervix when the edge still lay over it. Thanks to ultrasound, however, the location of the whole placenta can be much better located—even when it is behind the baby, i.e., closer to the mother's spine.

In the 1970s, improved computation, thanks to integrated circuit technology, saw the introduction of real-time imaging. Today's transducers send and receive millions of pulses each second.[2] Much smaller probes are now used, and a typical desktop computer has replaced the room-sized machines from back in the day. Although ultrasound was initially successful in imaging fetuses around twelve weeks into gestation and older, the use of probes in the early 1970s opened the possibility of detecting heartbeats and the imaging of fetuses around eight weeks in gestation via vaginal scanners.[3] These developments were made possible thanks to the integration of ultrasound imaging with a special property of waves that has had a particularly powerful application in astronomy and later in obstetrics.

Christian Andreas Doppler, who would become responsible for laying the foundation of what would become modern ultrasound technology, had some brilliant ideas that transformed physics. Born in the Austrian city of Salzburg on November 29, 1803, Doppler studied philosophy at the Salzburg Lyceum before going to the University of Vienna to study higher mathematics, mechanics, and astronomy. After teaching at the Technical Secondary School in Prague in 1837, he moved to the Technical Institute in Prague where he became increasingly fascinated by astronomy. Doppler became deeply interested in how the earth's motion could affect the color of light that came from a star and how a star's motion might be large enough to cause color shifts in the light that it emits. He described how the color of stars depends on the frequency of the incoming light that is measured and postulated that when a source moves away or toward an observer, then the frequency of the light shifts.

Doppler presented these findings—"On the Colored Light of Binary Stars and Other Stars of the Heavens"*—to the Royal Bohemian Society of Science in Prague on May, 25, 1842. In the work, which was improved upon four years later, Doppler postulated that the natural color of stars is white or slightly yellow, but when a star moves toward the observer, the wavelength of the light

* The original German being: *Über das farbige Licht der Doppelsterne und einiger anderer Gestirne des Himmels.*

would shrink and the frequency would increase, turning from green to blue, violet, and invisible (ultraviolet). A star that is moving away, however, would turn from yellow to orange, red, and then invisible (infrared)—its wavelength stretching and frequency decreasing. Although such changes in the colors of stars were impossible to observe with astronomical instruments at the time, Doppler realized that his idea could eventually become a powerful method in astronomy. "It is almost to be accepted with certainty that this will in the not-too-distant future offer astronomers a welcome means to determine the movements and distances of such stars," he wrote.[4] Doppler was certainly correct regarding its use in astronomy, and today it is employed, for example, to detect planets outside our solar system, but his theory was not just about light. It could also be used for any kind of wave, including all other electromagnetic waves such as X-rays, as well as sound waves.

Doppler's work was also not without its critics, and one of the biggest was the Dutch mathematician Christoph Hendrik Diederik Buys Ballot. In June 1845, Buys Ballot designed an experiment to refute Doppler's theory by using horn players from a brass band and getting them to play a single note as they traveled on a train between Utrecht and Amsterdam.[5] Buys Ballot positioned himself, together with some trained musicians, on a station platform and listened carefully as the train approached and receded. They found that the note played was indeed higher when the train approached and lower when the train receded. Just as Doppler described, its frequency was changing. The funny thing is that by attempting to prove Doppler wrong, Buys Ballot actually proved him right. Even with such an emphatic result, however, Bays Ballot still refused to accept Doppler's theory. You can carry out your own test next time you hear the sirens of an emergency service vehicle. Close your eyes and listen carefully, and you may notice the pitch change as the police car or ambulance approaches, the frequency of the sound increasing as it approaches and decreasing when it moves away.

While astronomy was a clear beneficiary of the Doppler effect, it took many years before it led to other applications. Today, for example, it can be used to catch you speeding on the highway. The first medical use that took

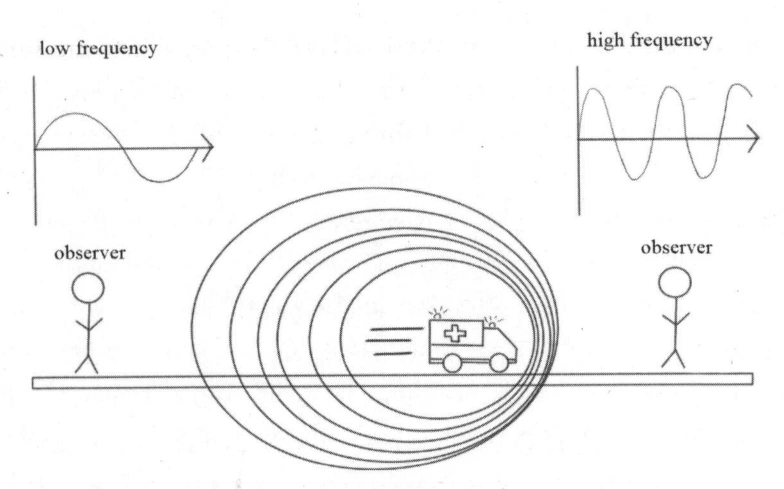

Fig 1 | *Schematic of the Doppler effect. As a source moves away from an observer (left), the frequency of the waves decreases, while they increase in frequency as the source moves toward an observer (right).*

advantage of the Doppler effect came in the 1950s to investigate the heart.[6] Then, in the later 1950s, biophysicists and doctors working at the University of Washington created a device to detect fetal heartbeats.[7] This was a handheld device that contained a piezoelectric transmitter and receiver. Any movement of blood toward or away from the transmitted signal would reflect the signal, with the frequency slightly altered, thanks to the Doppler effect. Blood moving toward the probe would result in a higher frequency, while blood moving away would lower the frequency. How much the frequency changes depends on how fast the blood is moving. This difference between the transmitted and reflected signal could then be amplified and filtered and either shown on a screen in "Doppler mode" or heard through earphones or speakers.

The 1970s saw researchers use both ultrasound and Doppler measurements to study the blood flow to and from the umbilical cord for the first time.[8] The following decade brought another computational advance with the introduction of 3D/4D ultrasound imaging and more precise measurements of the intricate blood flow in the fetal heart. The Doppler effect is also behind handheld probes you can use at home to detect the fetal heartbeat (despite advice against doing so from doctors and the US Food and Drug Administration).[9]

Today, ultrasound is an obstetrician's workhorse, able to measure every aspect of the fetus quickly and safely—from the size of the bones, head, and organs as well as the blood flow through the umbilical cord, the placenta, and the heart. Given the versatility of ultrasound, the International Society of Ultrasound in Obstetrics and Gynecology recommends that pregnant women have an ultrasound between eleven weeks' and thirteen weeks six days' gestation—called the "nuchal scan"—in countries that have the resources to perform them.[10] This involves basic measurements of the size of the baby to give an estimate of the "due date." The society also recommends that pregnant women have an additional ultrasound—called the "anatomy scan"—between eighteen weeks' and twenty-two weeks' gestation, which is when more detailed measurements are taken of the heart, placenta, and other organs. If any problems are spotted in the ultrasound exam, such as with the heart, then further tests can be carried out and plans made so that potentially lifesaving treatment can be given right after birth.*

As well as the medical benefits, the other valuable aspect of ultrasound is, of course, being able to connect with your little one for the first time and watching him or her kick and move around in the uterus, heart beating away on-screen. For all that, you can thank several developments in physics that began over 150 years ago.

---

* Congenital heart conditions affect up to eight in every thousand babies born in the United Kingdom: https://www.nhs.uk/conditions/congenital-heart-disease/.

BABY BUMP AND BODY IMBALANCES

One of the most exciting and precious aspects of pregnancy—after seeing the fetus on an ultrasound—is feeling a kick from the baby for the first time. Such instances are a chance to "connect" with your little one, like he or she is offering an outstretched hand, or more likely, foot.

Around week seventeen in gestation is when the earliest kicks may be felt, but identifying them can be difficult. Impacts around this period, Claire reliably informed me, felt like "flutters," the passing of gas, or little bubbles popping inside the stomach. She did not realize it during the first pregnancy, perhaps thinking a dodgy curry was responsible. But the second time around, she was much more aware that it was perhaps the baby moving. From week twenty onward is when kicks are more readily identified, becoming more and more vicious until they are not only felt but clearly seen from the outside—strong enough to cause the skin in the abdomen to eerily rise. I recall it being somewhat reminiscent of the "chestburster" scene in *Alien* (although not in the chest and without all the blood).

It is remarkable to think that the first movement a fetus makes in utero is at just seven weeks into gestation, which tends to be a "sideways bending" of the body. If you have an eight-week ultrasound scan, you may see a small figure jolting up and down, as if a pulse of electricity were coursing through its body. This time also marks the onset of a whole range of movements, with week nine bringing hiccups, the movement of an arm or leg, and sucking. Week ten, meanwhile, kicks off breathing, hand-face contact, and head rotation as well as opening the jaw followed by yawning from week eleven.[1] When the twelve-week ultrasound scan comes around, the fetus is well on its way to performing a daily routine of actions, much to the exasperation of the sonographer trying to catch that perfect shot for anatomical measurements.

Kicks or general movement can be annoying, especially during the night, and both our boys were especially good at this. Just as Claire was about to head to bed, it would turn into party time, leaving her attempting to get to sleep while suffering a good kicking. Women tend to report more fetal movement in the evening, but it is unknown whether activity is increased at that time or whether it is just because they are more aware of movements. After all, it is that part of the day that is usually spent sitting in front of the TV (enjoy it while you can!). Kicks tend to peak in the second trimester, which is roughly weeks thirteen to twenty-eight, and decrease in frequency in the third trimester (from weeks twenty-nine to forty) due to the reduced space in the uterus. These movements not only let the baby explore its environment but also, in the latter stages of pregnancy, perhaps create a "map" of their own bodies.[2]

While being kicked and punched is annoying at times, given that fetal movements are an indicator of health, it is also reassuring that the baby is happy and active. Studies show that around a quarter of women who perceived decreased fetal movements had either preterm births, low-birth-weight babies, or, sadly, a stillbirth.[3] Women are therefore encouraged to "count the kicks" during pregnancy, although this is somewhat of a misnomer. It is rather a change of movement pattern that should be monitored. This is a tricky thing to do, however, as we do not know what "normal" levels of movement are and why some fetuses move more than others—activity can vary from four to a hundred movements per hour. We also know little about fetal movements in twins and whether there are general gender differences.*

Another issue is perception. A 1980 study, which involved carrying out ultrasound scans on women in the third trimester of pregnancy, asked them to describe what they felt as they were being scanned. The researchers found that most women could accurately pinpoint large fetal movements but thought things like Braxton-Hicks contractions† were also the baby. On top of that, fetal movements that were easily spotted by ultrasound were not picked up by the mother at all.[4]

Biomedical engineer Niamh Nowlan at University College Dublin has been investigating the mechanics of fetal movements for over a decade. With a PhD from Trinity College Dublin, she began studying fetal chicks, focusing on the formation and conditioning of their bones—a process that can be studied rapidly, thanks to the twenty-one days that it takes a chick to go from embryo to hatching. Via optical imaging, Nowlan's research looked at how chicks develop complications in their joints from a lack of full movement in utero. In 2011, Nowlan moved to Imperial College London, where she turned her attention to applying those same techniques to examine the stresses and strains that are placed on fetal leg joints. To do so, however, she first had to get hold of images of babies kicking in utero, not an easy thing to capture.

Fortunately, a few years later, a £10 million (about $13 million) European initiative called the Developing Human Connectome Project (dHCP) began taking hundreds of MRI and other imaging scans of fetuses between twenty to forty-four weeks' gestation. The main purpose of the ongoing initiative is to study human brain development. Nowlan thought she could use this treasure trove of data to also study the biomechanics of fetal kicking. As they trawled through the data, Nowlan and colleagues found over three

* A small study of thirty-seven fetuses in 2001 showed a difference in male and female movements in late gestation, with males producing more leg movements.

† Braxton-Hicks contractions, commonly felt during the second or third trimester, are considered false labor pains.

hundred instances of kicks in action, which showed a clear extension of the leg—exactly what they were looking for. It allowed the team to deduce, for the first time, how much force that little David Beckham or Megan Rapinoe could produce in the womb.

If you ask somebody to name a law in physics, it will probably either be Einstein's energy-mass equivalence equation, $E = mc^2$, or perhaps Newton's second law of motion, $F = ma$. The latter, one of the first physics equations you likely came across in school, was formulated by the seventeenth-century English natural philosopher Sir Isaac Newton. Regarded as one of the greatest scientists to have ever lived, Newton is best known for his theory of gravitation, which, as the legend goes, originated after he was hit on the head by a falling apple as he sat under a tree at Woolsthorpe Manor, where he grew up.

Newton made significant contributions to several areas of physics, from optics and fluid mechanics to the foundation of classical mechanics via his famous three laws of motion, which he stated in perhaps the most famous work in physics: *Philosophiae Naturalis Principia Mathematica*, known simply as *Principia*. The groundbreaking work, published in 1687, introduced the concept of force, which is an interaction that changes the motion of an object—like a push or a pull. Force has units of Newtons (N), and the easiest way to get a sense of 1 N is to simply hold a 100-gram apple (or whatever else with that mass) in your hand.*

To get an understanding of the forces at work when the fetus is kicking, Nowlan, together with Stefan Verbruggen and their colleagues, created a computer model of the uterus and infant leg using a "finite element" method. This is a powerful technique in engineering that can also simulate how heat diffuses in an object or fluid flows through a pipe. It typically involves creating a mesh, consisting of thousands of small elements that form the shape of the

* Using Newton's second law of motion, or $F = ma$, with a mass (m) of 0.1 kilograms and acceleration due to gravity (a) being 9.81 meters per second squared, results in F = 0.1x9.81 = ~1 N.

structure, be it a teapot or a pelvis. The trick with this kind of analysis is that the calculations are made for every single element and then combined to give the result for the whole structure.

Nowlan's model of fetal kicks involved a finite-element simulation of the uterine wall, which is about 6 millimeters thick and shaped like a semicircle. A thin semicircle layer was added inside the arc of the uterus to emulate the amniotic sac, which the baby would first impact when it lashes out. This sac consists of two membranes. First is the amnion that contains the fetus and the amniotic fluid, the liquid that surrounds the fetus. Then there is the chorion, the outermost membrane that encases the amnion and becomes an integral part of the placenta.

To deduce the kicking force, the researchers had to consider the elastic properties, or stiffness, of the two layers. A stiff layer would take a greater force to deflect than something that is more flexible. A measure of the stiffness of a solid material is Young's modulus, which describes the ratio of stress, or the force per unit area, and strain, which is the deformation of a solid due to stress.* The team used Young's modulus values for the chorion and amnion, taken from experimental data: the chorion is similar to foam and the amnion is stiffer, more like rubber.

Using the dHCP images, the team measured how far the fetuses could deflect the uterine wall, and with the model, simulated a baby's foot with a force necessary to produce that same level of deflection in the uterine wall.[5] Their work showed that at twenty weeks' gestation, the kick matched a force of 29 N—about the same as that a two-year-old child can exert when pushing with a thumb. Carrying out the same analysis for fetal kicks at different gestational ages, the researchers discovered that the kicking force almost doubled to 47 N at thirty weeks, before dropping down significantly to 17 N at thirty-five weeks. They also found that this action could deflect the uterine wall by around 1 centimeter between twenty to thirty weeks, before dropping to just

* Young's modulus was developed by the nineteenth-century British scientist Thomas Young.

4 millimeters at thirty-five weeks—giving you a ten-week window to film that alien-like movement in action.

The simple reason for the decrease in force and deflection at thirty-five weeks is because there is less room in the uterus for the baby to extend its leg fully. You might think that is the end of the matter, but it turns out that this period is perhaps even more crucial for the baby's development than practicing all those leg extensions. The team next carefully carried out a 3D finite-element reconstruction of the fetal pelvis and the two main bones in the leg: the femur (thighbone) and tibia (shinbone). They simulated the stress and strain in the joints as the baby moved the foot against the uterine wall.

They discovered that while the force the fetus imparts on the uterine wall decreases after thirty weeks, from week thirty to full gestation the strain on the joints increases, especially at the joint between the tibia and femur. This straining of the joints is like doing resistance training in the gym. The problem is that while full-term babies get the benefit of the "utero gym," the same cannot be said for premature babies that often end up spending most of the time on their backs in neonatal wards, without the benefit of leg conditioning.

Nowlan says that this lack of conditioning could result in problems developing later. One high-risk disorder for infants that do not move much in utero is developmental dysplasia of the hip, which affects about one in every thousand births. It happens when the pelvis and the femur—or the "ball and socket" joint of the hip—fail to connect properly. This is more prevalent in firstborns, possibly due to the uterine wall being slightly stiffer the first-time around. Babies are also more at risk from the condition if they were in the breech position during gestation, that is, having their feet rather than their head at the cervix, which sometimes requires caesarean section due to possible birth complications. In this case, fetal movements tend to be different from non-breech positions and are also generally reduced.

To assess the risk of developing such conditions, which sometimes require surgery or the wearing of a harness around the hip for weeks, Nowlan thinks that an analysis of kicking ability could be performed during routine ultrasound scans. Despite the potential for longer scan durations to catch a kick in action, and therefore added cost, a program could be used to calculate the kicking forces automatically from the images. Detecting any anomalies early could result in better outcomes in the long run, with newborns potentially having light-touch physio programs to stimulate what they had missed in the uterus. Nowlan cautions, however, that this would have to be carefully done to avoid fractures in the weak bones of premature babies.

Armed with a knowledge of fetal kicking forces, Nowlan and colleagues are now investigating novel ways to detect the amount of fetal movement in utero. Having a hospital ultrasound to check on movement, while better than nothing, only offers a snapshot of how much the baby is moving. Techniques that could offer longer duration measurements, like cardiotocography, which measures fetal heartbeats, on the other hand, are not sensitive enough to accurately determine movement. Collaborating with mechanical engineer Ravi Vaidyanathan from Imperial College London, Nowlan's team has designed a wearable belt-like system that consists of eight small acoustic sensors placed equally apart in a circle on the abdomen.[6] Each sensor contains a diaphragm covering a sealed chamber that holds a microphone. When the fetus moves, a low-frequency vibration is produced that travels through the uterus and into the membrane. The resulting pressure change in the chamber is then picked up by the microphones and detected as a movement. To discern those movements from maternal ones, the system contains an accelerometer, similar to what your mobile phone detects when you move around.

The system has been piloted in forty-four women who were between twenty-four and thirty-six weeks' pregnant, collecting a total of fifteen hours of monitoring. The women wore the device while undergoing an ultrasound, with the system aiming to detect three types of fetal movement: quick, jolt-like startles; breathing; and slower whole-body movement. The system was particularly good at picking out startles—movements that the mothers are likely to feel—doing so 78 percent of the time, but less so for general whole-body movements (53 percent) and breathing (41 percent). Although the technique could discern between fetal and maternal movements, it cannot currently be used in the real world given the lack of sensitivity to certain movements and, therefore, potentially giving parents a false warning. The findings show, however, that acoustic sensors, or similar technologies, could one day be used to safely detect the frequency of fetal movements over longer periods of time. If it were possible to make a system more reliable with further improvements, as Nowlan is now planning, it could be used as an early-warning sign for reduced movement that could then be further investigated in the hospital.

As a father-to-be, I didn't have the luxury of knowing exactly what it feels like to be kicked in the womb, but I came much closer to learning how it feels to carry the literal weight of pregnancy. When Claire was six months pregnant, we went to a local baby superstore to spend all our hard-earned money on things we thought we would need. While deciding whether to go for the blue or white baby mittens, we were approached by two shop assistants who, as the saying goes, must have seen us coming a mile away. The two women were holding what looked like a strange backpack. It was black with a big bulge and a couple of arm straps. "Would you like to try it on?" one of them asked me. I didn't really know what to say, so I just nodded rather hesitantly. I put my arms through the straps, adjusted it into place, and then tightened the straps so it fit snuggly over my coat. It soon became clear that it was a "pregnancy simulator," a heavy "frontpack" to mimic the average added size and weight of carrying a baby to full term. At first, it did not feel too heavy; I bent back slightly to accommodate the extra mass and then started to waddle around the store. Off we went to look at some bottle teats while getting second glances from people who thought I had taken pregnancy sympathy to another level. I only had the simulator on for around ten minutes, but the more I walked, the heavier it felt. By the time I returned to hand it back, I whipped it off as quickly as I could before carrying on shopping for overpriced bibs.

In that instant, my body gained an extra 11 kilograms (about 24 pounds), which, thankfully, is not the same during pregnancy where weight gain is slower coming, giving the body the chance to adapt. Indeed, many women take the "side-on selfie" during pregnancy to monitor how the baby bump is progressing week by week. With the pregnant uterus beginning its journey in the pelvis during the first trimester,* initially there is no noticeable difference, even when carefully lining up the exact same frame for each week's shot. That begins to change, however, around twenty weeks' gestation, when the uterus moves into the abdomen and a slight baby bump starts to poke out. This normally marks the point when it becomes a bit harder to hide the news. When a bump becomes more prominent, the size of the bump in centimeters corresponds roughly to the week of the pregnancy. (The explanation for why this numerical equivalence exists is beyond modern science.) At thirty-weeks' pregnant, for example, the fundal measurement—the length from the top of the uterus to the top of the pubic symphysis†—will be around 30 centimeters (close to 12 inches).

All that growth over a relatively short time in utero still places a significant energy demand on the body. That little bundle of cells, as outlined in chapter three, starts off around 0.1 millimeters in diameter (about the size of the period at the end of this sentence) but nine months later turns into a 3.5-kilogram (about 7.7-pound) behemoth that gains around 200 grams (about seven ounces) each week by the end of pregnancy. This weight gain takes a huge amount of energy, despite the overly used and inaccurate term "eating for two."‡ In 2019, researchers at Duke University discovered that, during pregnancy, a woman's energy use peaks at 2.2 times her resting metabolic rate, or the amount of energy a person needs to keep the body functioning at rest.[7]

It is thought that human energy expenditure cannot be sustained above two and a half times the body's resting metabolic rate, as at that point the body is burning calories quicker than it can absorb from food, representing a ceiling of performance. Examples of this ultimate energy demand include the Tour de France or doing an ultramarathon. Carrying a baby, then, is not far

* Running from conception to week twelve of pregnancy.

† The pubic symphysis is a joint between the left and right pubic bones.

‡ It is recommended that pregnant women only need an extra 200 additional calories in the third trimester of pregnancy.

from an endurance race, and the research suggests that these same physiological limits that keep Ironman triathletes from continuously breaking records could be the same factors that constrain how large babies can grow in the womb. Yet, despite that apparent ceiling, carrying a baby still takes a heavy toll on the body. And there comes a point, well into the third trimester, when the weight and size of the baby begins to affect a woman's stability.

In 2009, the US Institute of Medicine recommended that women with a normal pre-pregnancy body mass index, or BMI, gain from between 11.3 to 15.9 kilograms (about 25 to 35 pounds), or add five points in BMI, during pregnancy.[8] Most, but not all, of this is centered on the abdomen, which increases by around 31 percent, or 6.8 kilograms (15 pounds). With such a large mass protruding from the abdomen—and with no possibility for shoulder straps—it is amazing that you do not see heavily pregnant women falling over all the time as they go about their business. And such stumbles can be a serious issue: falls are the number one cause of traumatic injury during pregnancy that can lead to premature birth.[9] So, how do most pregnant women manage to keep upright all the time when moving around?

Women experience many forces when carrying a baby, and not just from the fetus kicking and punching. Gravity and the weight of the baby are the biggest forces to consider. Bear with me on this one, but the explanation for why heavily pregnant women *could* struggle to keep upright are two concepts that are closely related to force: torque and center of mass. Torque is the rotational effect caused by a force. An easy way to think about it is to try shutting a door. It is much easier to shut a door using the handle than it is by pushing it nearer the hinge, or where the moment of rotation occurs.* Obviously, the door handle is placed there for a reason, and part of that is because it takes less effort to shut it the farther away you are from the hinge.

---

* If you take the analogy to its extreme, it is not possible to shut a door by applying a force at the hinge.

A key aspect for thinking about torque in a body is to consider its center of mass. All the way back in 250 BC, the Greek mathematician Archimedes—who deduced relation between a circle's circumference and diameter, known as π—showed that the torque exerted on a lever by a collection of weights at various points along the lever is the same as if all the weights were moved to the center of mass. Determining the center of mass of a 2D object can be thought of as the point where you need to support an object to make it balance. Imagine balancing a pencil with your finger. If you do it near one end (assuming it is homogenous), it will fall by rotating around your finger, therefore having a torque. If you put your finger in the middle, you will be able to balance it—the center of mass being in the middle. Of course, many objects are not so simple, and if a mass is unevenly distributed, say it is denser at one end, then the center of mass would shift toward the heavier part and you would need to put your finger farther along in that direction to balance it. The center of mass, therefore, is the imaginary point at which the whole mass of the object is concentrated.

Some animals use center of mass and torque to their advantage. For chimpanzees, given that they are mostly quadrupedal, their center of mass is located between their front and hind legs, and is relatively low off the ground. This has its advantages when carrying an infant. The baby grows roughly where the center of mass is, meaning that as the infant adds mass, the mother's stability is not greatly impacted in terms of her postural support base. For bipedalism, however, humans pay a price when it comes to balance. A human has a center of mass that is high off the ground, about 10 centimeters (4 inches) lower than the belly button, near the top of the hip, and around 0.5 centimeters (about two-tenths of an inch) forward from the hip. Given that the location of the center of mass is slightly above and in front of the hip, this results in a torque that is generally fine to stop us from constantly fighting the need to tip over, but adding a lot of mass that protrudes from the front of the body in such a short space of time, as in pregnancy, can pose issues.

In 2007, scientists at Harvard University and the University of Texas at Austin showed just how much a fetus and additional weight, such as the placenta, can affect the center of mass and also how the female body can compensate.[10] The study included nineteen women as they progressed throughout pregnancy. The researchers took measurements of the women's center of mass at six equally spaced times during gestation. The center of mass was calculated using computer modeling and a force plate, which is basically a set of large scales that measure ground-reaction forces in three dimensions, similar in some sense to a Nintendo Wii Fit Balance Board.

As the fetus grows in late pregnancy, the woman's center of mass starts to move forward from the hips, so that by full term it is estimated to be about 3.2 centimeters (1.25 inches) in front of the hips. This would result in a torque around the hips by the upper body about eight times greater than that of a nonpregnant woman—likely resulting in a lot of imbalance and potential falls. The researchers found, however, that some women seemingly let the center of mass move forward, but only until a certain point, which is usually around week thirty in pregnancy, or when the fetus is about 40 percent of its final total mass.

To find out what was happening after this point, the team analyzed the position of individual vertebra in the lower spine by sticking reflective markers at the vertebral positions on the body and then tracking the reflectors with infrared cameras as the women stood. The human spine has a natural forward curve at the bottom that helps to compensate for the extra pressure of walking on two feet. The lower vertebrae in women are also more wedge shaped compared to the squarer shape in men, which allows women to curve the lower back over three vertebrae as opposed to two in men. Beginning around week thirty in pregnancy, some women naturally bring their center of mass back to compensate for the additional mass, but this requires a truly back-bending move called gestational lordosis.*

The team found that the angle of gestational lordosis—an angle between four vertebra in the lower spine as it curves—is about 50 degrees in a pregnant

* It has also been found that not all women do this backbreaking move and they still manage to carry the baby to term, although it is not known what changes happen to facilitate these cases.

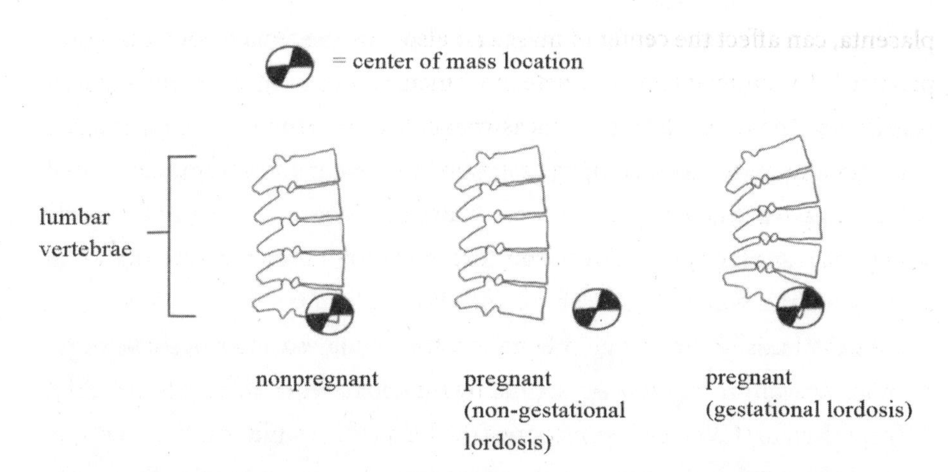

Fig 1 | *The body's center of mass lies around the bottom of the lower vertebrae (left). During the third trimester of pregnancy, the added mass from the fetus and placenta can cause the body's center of mass to move forward from the hips (center). A way to counteract this is gestational lordosis, in which the lower vertebrae bend (right) to bring the center of mass back to the posterior.*

woman at full term, compared to 30 degrees for a nonpregnant woman. A higher angle results in a larger "bend" in the lower spine and that helps to bring back the center of mass to almost where it was pre-pregnancy, resulting in a small torque. Yet, this move comes at a cost through increased shearing forces on the spine that can cause back pain. However, the researchers found that the joints between each vertebra are larger in women than in men. This enlargement helps avert serious injury, notably preventing the vertebrae from painfully slipping sideways as they move. While the mechanism for this back-bending move is not fully understood, the hormone relaxin, produced by the ovaries and placenta, could be a significant contributor. Concentration of relaxin peaks around the twelfth week of pregnancy and provides greater ligament laxity in the pelvis and joints.[11]

Robert Catena at Washington State University has been studying balance for over a decade, initially focusing on changes following injury, especially occupational injuries. When he moved to Washington State in 2014, he turned his attention to studying the gait of pregnant women. Catena suspects that the shift in the center of mass in later pregnancy due to gestational lordosis is maintained during walking and is not just a stationary effect as determined in 2007. He built an "anthropometric" model, which he hopes will answer the dynamic gestational lordosis question. His model breaks the body up into around fifteen segments so that the center of mass can be determined for each segment as well as the whole body. Dynamic posture can then be assessed using motion-tracking software that takes measurements of about fifty points on the body as the person moves across a force plate.[12]

Every part of the body, from the thighs to the arms and feet—not just the torso—adds mass during pregnancy. Catena's work shows that taking a segmented view of the pregnant human body can highlight why some tasks can cause imbalance in pregnant women. For example, with an increased arm mass, simple things like reaching for an item in the supermarket could affect balance due to an increased and not-accustomed-to torque from this section of the body. But, in some sense, worrying about weight gain alone is also a bit of a misnomer. After all, the incidence of falls increases rapidly in the second trimester, when the added mass of the fetus is not as high, and then plateaus in the third trimester.[13] The reason for this could be because pregnant women tend to move less in the latter part of pregnancy. It could also be that as pregnancy progresses, the body adjusts its posture or gait for more stability that is in line with the center-of-mass shift. Some of these changes involve an increase in step width, or the horizontal distance between the two feet when walking, as well as a decrease in stride length. This combination not only better supports pregnant women but also enables each foot to stay on the ground for a longer duration when walking.

Overall, Catena has found that extra weight cannot fully explain balance change, showing that other issues must be at play.[14] He suspects that aspects such as a decline in vision, attention changes, and muscle fatigue are likely to have just as big an impact as the back-bending gestational lordosis, which he has shown can occur to various degrees in women and could perhaps be more pronounced for shorter women.[15] "It's not all mechanical, but likely psychological too, given that women focus more on their bodies during pregnancy,"

Catena explained, "which, of course, is good, but it comes as a detriment to being aware of their environment."

According to Catena, obstetricians at routine pregnancy appointments rarely ask pregnant women whether they have had any falls or balance issues that could then potentially be investigated further. With that in mind, he wants to create a clinical tool—one that would be more objective than being solely based on a simple questionnaire—to estimate the risk of losing balance. Although a clinical setting would not have motion-capture capabilities and center-of-mass modeling, such a tool suggested by Catena could consist of a few standing balance exercises that are then extrapolated to dynamic balance via computer simulations. Catena added that the most important aspect of his ongoing work in this area is determining the critical factors that impact balance in pregnant women, whether mechanical, physiological, or psychological.

"There is no other population out there that experiences such naturally drastic changes in such a short period of time," he told me. "Ultimately, this is about understanding human pregnancy as a limiting factor in bipedalism."

6

READY, SET, (CONTR)ACTION!

If your only knowledge of what happens when labor commences is from TV or at the movies, then you will probably think that the "water breaks" most likely when out grocery shopping, resulting in a gush of amniotic fluid on the floor near the fruit and vegetables aisle. The next scene includes a few screaming pushes in the hospital, and before you know it, the happy couple is cradling a newborn with just a slight glistening of sweat on the forehead.

Although labor can happen quickly, this order and speed of events is far from the norm. Studies have shown that the rupturing of the amniotic sac before the onset of contractions occurs in only about 10 percent of pregnancies.[1] For many women, their water does not break until they are in labor. For both our children, my wife did belong to this 10 percent group, but thankfully her water broke in the early hours of the morning at home (and, yes, we did have a mattress protector). It was far from a gush—more like a trickle, followed by a few more hours waiting for the contractions to begin. Some twelve hours later, the baby was born. What I took from that experience is that nothing happens like it does in the movies.

I have never felt contractions and have no desire to try one of those contraction simulators, which sounds like it should be a CIA torture technique.

My wife describes them as feeling like all your innards are being squeezed all at once—similar to terrible stomach cramps caused by food poisoning. I have, however, witnessed the true force that contractions can produce. When I was watching, or I should say helping, my wife in the latter stages of labor with our second born, she was in the birthing pool to take away the heavy, achy pain when a huge contraction seemed to lift her entire abdomen like a pulsation. It was almost like the baby was picked up and forced down toward the hips, an incredible sight that highlighted what the human body is capable of. Although, I cannot be 100 percent sure it was not some light refraction effect or perhaps my tired eyes playing tricks on me at 3 AM.

All this pushing power comes from the uterus, which is shaped like a 7.5-centimeter-long (almost-3-inch) upside-down pear. At the neck of the uterus is the cervix, while the top is the fundus. The uterus changes rapidly during pregnancy, with the thickness of the uterine wall staying constant until around sixteen weeks' gestation before beginning to thin and elongate as the fetus grows. The uterus grows to such an extent during pregnancy that it becomes one of the largest muscles in the body, with a mass of about 1.3 kilograms (close to 3 pounds) at full term compared to just 75 grams (about 2.6 ounces) before pregnancy.[2] It can also hold around 5 liters (about 21 cups) of fluid at full term compared to just 10 milliliters (about a third of a fluid ounce) in the nonpregnant uterus.

The job of these powerful contractions is to use the baby's head as a battering ram to shorten the cervix and then to push the baby through. Pre-pregnancy, the cervix is stiff and tough, while the pregnant cervix, thanks to hormonal effects, is much softer, like a rubber band. This "cervical ripening" is the last step before labor, when the cervix then dilates to allow the baby to pass through the birth canal.* One way to visualize what these contractions are doing to the uterus and cervix is to get a Ping-Pong ball and put it inside a deflated balloon. Inflate the balloon and make sure the Ping-Pong ball rests at the bottom over the neck of the opening. The balloon represents the uterus,

* After birth it takes about six weeks for the uterus to return to its nonpregnant size and for the cervix to close back up, on average.

the ball is the baby's head, and the neck of the balloon is the cervix. Now, put both your hands on the top of the inflated balloon and gently push for a brief time before relaxing—like a contraction. As you carry this out multiple times, you will start to see more and more of the ball at the neck of the balloon as it "dilates," until eventually the ball pops out and is "born."*

When my wife and I were doing our antenatal classes, we were told of many activities that *could* kick-start labor. This included drinking raspberry tea, going for long waddle-like walks (without falling over), having sex, or eating spicy food. Sorry to spoil the fun, but these are all myths and hearsay. None of this will help. According to science, however, there is one thing that could work: nipple stimulation. An analysis in 2005 of four separate studies, which randomly assigned each full-term pregnant woman to either a "stimulation" or "non-stimulation" group, found that about 37 percent of those who stimulated their breasts entered labor within seventy-two hours compared to just 6 percent in the non-stimulation group.[3] Before you start thinking that this is the answer to your about-to-go-overdue prayers, there is a catch: it requires dedication. Women had to stimulate their breasts between one to three hours each day. And, of course, not everyone who did so went into labor. So, after all that effort, it might not work anyway.

The reason why nipple stimulation can have an effect is because it causes a chain of events that results in the uterus contracting, which can induce labor. Rubbing or rolling the nipples helps to produce a special hormone called oxytocin,† which is released into the bloodstream by the pituitary gland, a pea-sized structure at the base of the brain. When the hormone makes its way to the uterus, it also stimulates the production of prostaglandins,‡ which in turn help foster contractions. We know that oxytocin has a powerful effect that can

* For a video of the balloon and Ping-Pong demonstration, see https://www.youtube.com/watch?v=q_5XWgEBQQA.

† Oxytocin is often referred to as the "love hormone."

‡ Prostaglandins are molecules that are produced in the body's cells and have a range of functions from repairing damaged tissue to triggering the uterus to contract. Manufactured forms of prostaglandins are used to induce labor.

kick-start and strengthen labor, with doctors administering it via injection to induce labor or quickly deliver the placenta following delivery of the baby.*

Once labor kicks in, the uterus transforms from being a dormant muscular organ for almost all an adult's life, producing infrequent, uncoordinated, localized, and ineffective contractions, to suddenly generating strong, rhythmic, and coordinated, contractions. But labor can also happen before it is supposed to. According to the US Centers for Disease Control and Prevention, around 10 percent of births in the United States are preterm—those that occur before thirty-seven weeks' gestation—most of which are due to either problems with the cervix or the early onset of contractions.[4] Preterm birth is a leading cause of death among children under five years of age, responsible for around a million deaths per year worldwide.[5]

Despite knowing the powerful stimulus that hormones have, we still have no idea why contractions begin when and where they do. And another mystery is how the uterus manages to coordinate this behavior across the whole organ. The best model we have for how a muscle undergoes powerful, coherent contractions is the heart, which uses special "pacemaker" cells to coordinate the rhythmic pumping of blood around the body. Could this explain in a similar way how the uterus manages to expel a 4-kilogram (8.8-pound) baby?

The human heart plays a crucial role in keeping you and me alive, and special attention is given to the organ whenever you go for a medical checkup. Likewise, detecting a fetal heartbeat at the eight- or twelve-week ultrasound scan is a major milestone of pregnancy. A positive result and you have the confidence to shout it from the rooftops. Although things can, and do, go wrong from that point on, seeing a fetal heartbeat at eight weeks means there is a 98.5 percent chance of having a "live birth."[6] At twelve weeks it is about a

* Some have claimed that oxytocin is not required to deliver a baby, pointing to evidence that in 2014 a brain-dead woman was able to naturally deliver a baby; see Kinoshita, Y; Kamohara, H., Kotera, A., et al. "Healthy Baby Delivered Vaginally from a Brain-Dead Mother." *Acute Medicine & Surgery* 2, no.3 (2014): 211–213.

percent better, which is usually when expectant parents start announcing the news to their parents, giving them six months to prepare for all the babysitting duties they will be doing. The fetal heart makes its first "beat" around six weeks into gestation, although at this point it is not a fully formed heart—that takes another four to six weeks. Rather, what you are seeing at an eight-week scan is a little flutter in the area that will become the baby's heart. This flutter is produced by a group of cells that will become the future "pacemaker" of the heart—having the ability to fire periodic electrical signals that contract heart-muscle cells called cardiomyocytes.

The heart's pacemaker region was first discovered by the British physiologists Arthur Keith and Martin Flack in 1907 in the hearts of moles, in which a special part of the organ—the sinus node—was found to be responsible for the beating.[7] Similar zones were later found in other mammals, including the sinoatrial node of the human heart, which is located on the back wall of the right atrium.

What these special heart cells in the node produce is an "action potential," an electrical signal caused by the rapid rise and fall of the cell membrane potential. This potential is related to the difference in concentration of several different ions across a membrane, which spreads through the muscle cells, causing them to contract, pushing blood through the ventricles. This action potential is achieved via an intricate dance of these ions in and out of the individual heart cells that results in voltage spikes, which transmit ions to neighboring cells.* When ions (such as potassium, sodium, and calcium—all positively charged ions) enter the cell, the voltage increases. When they flow out of the cell, the voltage decreases. Almost all cells in the body have "pumps" that control how fast ions cross the cell membrane via specialized ion gates—proteins that open and close pores in the membrane to allow ions to pass through. The crucial thing is that various ionic gates only begin to operate when the voltage of the cell crosses specific "threshold" values.

---

* Ions are charged atoms or molecules. A negatively charged ion has more electrons (electrons being negatively charged) than protons while a positively charged ion has fewer electrons than protons.

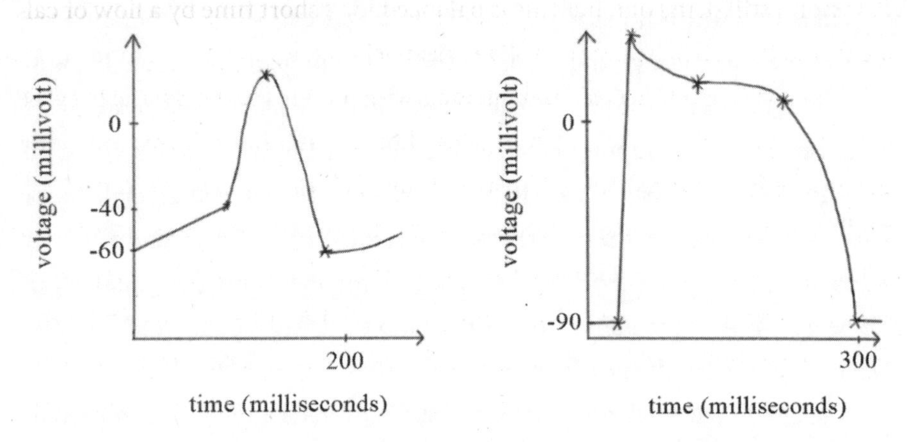

Fig 1 | *Schematic of the heart pacemaker action potential (left) and cardiomyocyte action potential. The (*) indicates different stages that result in certain ion gates opening to adjust the voltage.*

The voltage of a heart pacemaker cell starts off, let's say for the sake of simplicity, at around -60 millivolts*, at which point positive sodium ions enter the cell to increase the voltage to about -40 millivolts (see figure). When it reaches this threshold, calcium gates open, and the ion flows into the cell, raising the voltage abruptly to about +10 millivolts. Then, potassium channels open and they escape out of the cell, bringing the voltage down to -60 millivolts, and the cell relaxes, and then the cycle begins again. The result of this action potential is that calcium and sodium ions from the pacemaker are transported to the neighboring cardiomyocytes, which is done via protein complexes called gap junctions that connect the cells together.

The action potential for the heart muscle cells, or cardiomyocytes, is slightly different. They have a "resting potential" of about -90 millivolts where they remain until receiving a sodium and calcium boost via the pacemaker cell. When this happens, the cardiomyocytes' voltage rises to about -70 millivolts. Sodium gates then open and the ions flow into the cell, which raises the voltage incredibly quickly, causing it to overshoot zero millivolts. That, in turn, causes potassium to flow out again until the voltage drops again to zero.

---

* Voltage magnitude can be positive or negative.

Potassium still drifts out, but this is balanced for a short time by a flow of calcium into the cell, keeping the voltage relatively stable.

It is at this point that calcium interacts with proteins in the muscle to cause it to contract. When the calcium gates shut around 200 milliseconds after opening, potassium continues to leak out until the voltage is back again at -90 millivolts, and so the cycle starts again when stimulated by another pacemaker action potential. The pacemaker signal to the neighboring cardiomyocytes starts a chain reaction, like dominoes falling across the heart. The rhythmic impulses generated by the pacemaker cells directly control the heart rate, sending action potentials around seventy times per minute to keep the heart rhythmically pumping blood around the body for decades. If you have an average of eighty beats per minute over an eighty-year life span, then a simple calculation reveals that your heart would have made over three billion beats in a lifetime.

The heart is not the only organ that uses pacemaker cells. They are also in the smooth muscle of the intestine, where cells with the rather cumbersome name "interstitial cells of Cajal" produce an action potential that sends calcium into the smooth muscle cells of the intestine. This results in a contraction that propagates like a wave, aiding balls of food down through the digestive tract. Given that pacemaker regions are in multiple organs in the human body, it had long been thought that contractions in the uterus must be similarly coordinated with pacemaker cells switching on somehow during the end of pregnancy. Despite decades of painstaking work and much searching, however, no such cells have been found and, most likely, never will be. In some sense, this lack of a dedicated pacemaker zone makes sense for an organ that may never need to push out a baby, and even if it does, it would only need to be fully ramped up for a day or two in its entire existence. There are, however, some similarities between the muscle cells of the heart and the uterus.

Just like the cardiomyocytes, the muscle cells in the uterus—dubbed uterine myocytes*—are in some sense dormant thanks to an abundance of open potassium channels that result in a heavily depolarized cell. The heart's

* Myocytes connect to form fibrous structures, which in turn are organized into small bundles called fasciculi.

pacemaker helps to kick-start the muscles into action (literally), but this same effect doesn't happen in the uterus, meaning that there must be another source that can produce an action potential. Even if the uterus is stretched, for example being kicked by the fetus, which would result in the release of calcium, so-called "blocker" channels stop any synchronous behavior from establishing. While an oxytocin kick can produce an action potential, the uterine muscle cells are unable to spontaneously polarize and depolarize—or oscillate—like the pacemaker cells in the heart.

Searching for how the uterus contracts is made more difficult because other cells in the uterus, such as interstitial Cajal-like cells and fibroblasts, which play a key role in molding the uterus as it expands,[8] are electrically passive, so when they are stimulated by an electrical pulse, the signal quickly decays and fails to transmit to another cell. On top of that, without a pacemaker mechanism, the contractions somehow need to be coordinated for the right duration and frequency. If contractions are too long, then there is a danger of the baby becoming oxygen deprived. If contractions are too frequent, then delivery could be too quick, resulting in potential injury to the baby and mother. Having no pacemaker function means that the myocytes need to do something rather special to self-oscillate and synchronize this behavior coherently across the whole organ.

The fundamental rules that govern the dance of ions across a cell membrane and how they translate into an action potential in cells was first discovered by the British scientists Alan Lloyd Hodgkin and Andrew Fielding Huxley in 1952. The pair were working at the Physiological Laboratory in Cambridge and the Laboratory of the Marine Biological Association in Plymouth, not on heart cells, but the giant axon of a squid.[9] The giant axon (nerve fiber) is responsible for contracting the mantle muscle that propels the cephalopod through the water and is around a millimeter in length on average—possible to see with the naked eye.*

* This is similar to the high-school biology experiment in which a small voltage applied to a dead-frog muscle causes it to contract.

Hodgkin and Huxley measured the voltage through the axon via a technique called the "voltage-clamp" method. This involved carefully threading an electrode into the axon and, via delicate experiments in various ionic solutions, they showed that the nerve impulse is due to the voltage-controlled flow of sodium and potassium ions through the axon membrane.[10] From modeling these experiments, they came up with the Hodgkin–Huxley model, which is a set of four coupled mathematical equations that keep track of the voltage of the cell over time as the different ions flow in and out. It can accurately predict the main recognizable features of an action potential such as that seen in the heart cardiomyocytes: an upstroke, excited peak, refractory, and recovery phases.*

Despite the model's complexity, the work of Hodgkin and Huxley spurred much interest among other scientists. This included biophysicist Richard Fitzhugh who was working at the Biophysics Laboratory of the National Institutes of Health in Maryland. He began using an analog computer to find solutions to the Hodgkin–Huxley equations, and by doing so, formulated his own spin on the model. He took the main aspects of the Huxley–Hodgkin model and distilled it into a more basic one that had two just variables, as opposed to four in the Huxley–Hodgkin version, to produce a voltage spike and relaxation.[11] The oscillation in Fitzhugh's model—as in the case of the Hodgkin–Huxley model—is "nonlinear." A linear equation means there is a simple relationship between two variables that results in a straight line on a graph; as one variable gets bigger, so does the other. In a nonlinear system, however, the input and output are not proportional. In the case of an action potential, the voltage increases relatively slowly but then relaxes back down incredibly fast in a nonlinear way. This simpler version of the Hodgkin–Huxley equations, called the Fitzhugh–Nagumo† model, turned out to be of huge interest to scientists and led to the formation of a new field of applied

---

* For their work, Hodgkin and Huxley were awarded the 1963 Nobel Prize in Physiology or Medicine.

† Jin-Ichi Nagumo was a Japanese electrical engineer who built an experimental version of the model in the 1960s and thus put his name to it.

mathematics in excitable systems.[12] The work is still highly relevant today some seventy years on, especially for modeling neurons (see chapter twelve) and other excitable cells.

For the past decade, physicist Nicolas Garnier at the École Normale Supérieure in Lyon, France, and colleagues have been using this model to study the dynamics of uterine myocytes, especially how these cells could produce self-oscillations that translate into synchronized, coherent behavior. They were particularly intrigued by the role that interstitial Cajal-like cells and other passive cells could play. These Cajal-like cells alone contribute up to 18 percent of the cell population on the surface of the uterus, decreasing to around 8 percent in the myometrium.* The uterine myocytes have a resting potential of around -70 millivolts at the beginning of pregnancy, but that changes to about -50 millivolts midterm.[13]

In later pregnancy, thanks to oxytocin, calcium is kicked into the cell. There is a lot going on in the cell, but in a simplistic way, this calcium surge causes the voltage to spike (depolarize), which in turn causes potassium channels to open, repolarizing the cell once again to produce an action potential. That is fine for a single oscillation, but with no pacemaker, not enough to keep it going. Using the Fitzhugh–Nagumo model, Garnier and colleagues found that, on its own, a single myocyte failed to self-oscillate as a heart pacemaker could. It would increase in voltage from -50 millivolts to zero followed by an incredibly fast nonlinear decrease back down to -50 millivolts again without any further oscillation.

Garnier and colleagues then added passive cells to the model and connected one to a single myocyte. They found that the dynamics in this case were hugely different. This came about due to the difference in the resting potential of the myocyte and the passive cells. The interstitial Cajal-like cells, for example, have a resting potential of about -58 millivolts while the fibroblasts' is around -15 millivolts. In such a case, the myocyte would produce the

---

* The uterus consists of three main layers. The endometrium is the innermost layer that the embryo first implants into. The next layer is the myometrium, which contains the myocytes, while the third, outer layer is the perimetrium.

same single action potential, but because the passive cells have a lower resting potential, this lowers the resting potential of the myocyte-passive-cell system. What Garnier found, from a purely theoretical viewpoint, is that this coupled system does not have one "solution" in the Fitzhugh–Nagumo equations—as for the lone myocyte—but rather two.* This means that the system can flip between each solution, each time firing an action potential as it does so. In this case, the result of an oxytocin boost to the cells would not just generate a single action potential, but potentially several, each separated by a time that allows the cell to relax back to the resting potential.

Garnier and colleagues wanted to next understand the dynamics that would result from a collection of connected myocytes and passive cells. Experiments on rat uteri offer some clues about how this behavior can spread throughout the organ revealing that conductance, a measure of how easily electrical signals can pass through the cells, increases by a factor of about twenty in late pregnancy and labor. This huge conductivity boost is thought to be due to the opening of more gap junctions later in pregnancy that connect either the myocytes together or the myocytes with neighboring passive cells. Experiments in rats have shown that the number of gap junctions rises from about 50 per muscle cell in early pregnancy to 450 at full term.[14]

To see what effect the increase in conductivity or number of gap junctions had on the dynamics, the team created a 2D grid of 64-by-64 myocytes, with each one connected to either zero, one, or two passive cells in a random distribution. For weak coupling, or a limited number of gap junctions, the collective oscillations in the system from the myocytes were initially all out of sync. Some areas were excitable while others were not. But as the coupling increased, clusters of excitable elements began to oscillate with the same frequency, eventually resulting in the clusters merging to produce a single shared wavelike oscillation through the grid of myocytes. This showed that an increase in coupling not only leads to coherent activity but also that the excitations are initially irregular and then become more coherent toward the end of labor as the couplings increase.[15]

---

* Garnier says that there are actually three solutions, but the third is likely to be "unstable."

"When the conductivity is high between the myocyte and passive cells, the solution is not to relax to zero but to begin periodic wavelike oscillations," said Garnier, who expects that the increase in gap junctions could perhaps emerge from the effect of oxytocin. "Our model, at the least, could help to explain why this powerful drug works so effectively to induce labor."

It was a tantalizing result, but it was also a fairly simple analytical model of an active cell attached to passive cells. It certainly didn't replicate anything happening in the uterus. So, the team created a more realistic model of uterine myocytes, considering twenty different variables of the excitable cell, such as ionic currents for potassium, calcium, and sodium. "It took us about a year to work out what the relevant parameters were for this more realistic model," Garnier explained to me. "But it gave us a physiological model that we could test and compare to experiments." With this more complex model, they found, as before, that a myocyte could only self-oscillate if it was connected, or coupled, to a passive cell. They also found that if the uterus were only composed of interstitial Cajal-like cells and myocytes, this would not be enough to stimulate oscillations; the addition of fibroblasts was critical.[16]

When the team created a 50-by-50 grid of more realistic myocytes, they found some intriguing behavior, different than before—the onset of spiral waves through the grid. Spiral waves can occur in contractable organs, but it is usually not a good sign. In the heart, for example, they can cause irregular heartbeats known as arrhythmias. The problem with spiral waves in the uterus, however, is that by swirling as they move, they would contract in different directions, likely resulting in not much happening. Despite this, similar complex waves have been seen in the pregnant uteri of guinea pigs.[17] Indeed, when the team analyzed these spiral wave propagations as they passed through the grid, they saw that it took the wave about 340 milliseconds to do so. Given the length of the myocyte cells is about 225 micrometers, the speed corresponds to around 3.3 centimeters (1.3 inches) per second, which roughly agrees with the experiments on guinea pigs. They also found a coupling strength, or conductivity, between the myocytes of around 12 nanoseconds, corresponding to 240 gap junctions, which is, again, consistent with experimental findings.

Although this work shows that the passive cells and gap junctions are critical, not only in how the uterus contracts but also how it does so across the whole organ, the elephant in the (delivery) room is where the contractions begin in the first place. Successful delivery in humans is thought to be associated with strong contractions that begin in the fundus and pulsate down to the cervix (it was certainly what I saw during the delivery of my children), although this is not settled among scientists.

Garnier and colleagues modeled this fundal dominance by introducing a gradient of passives cells—a concentration of cells higher toward the fundus with the cell numbers gradually decreasing toward the cervix. When they ran their simulations, they found that for low coupling strength between the myocytes, localized small spiral waves appeared all over the system, like what they saw before. Once the coupling was increased, the waves began to propagate from the fundus toward the cervix, thanks to this difference in more passive cells at the fundus.[18] For now, this is just a hypothesis, and it is not known whether this is physiologically accurate, although Garnier points out that a study in 1999 found evidence for a higher concentration of passive cells in the upper part of the uterus.[19]

There are other possibilities for where contractions originate that are throwing up tantalizing results. At least, that is, in rats. Scientists in the United Kingdom, led by physiologist Andrew Blanks at the University of Warwick, have carried out mathematical modeling and detailed electromyographical recordings on the uteri of twenty-nine rats. This involved painstakingly taking slices of tissue—each a few millimeters thick—from the uterus, including the implantation site of the placenta, and assessing the excitations of the muscle via electrodes. Rather than solely taking electrical measurements in a 2D slice, as had been done before, the interdisciplinary team built up layers of the uterus to form a 3D model. Using mathematical modeling, they then discovered that most of the action potentials occur first in the smooth muscle cells close to the placenta before spreading out.[20]

When Blanks and colleagues then studied the muscle fibers at the interface between the placenta and uterus, they detected "bridge-like" muscle structures that protrude from the placenta toward the main smooth muscles in the uterus. Blanks thinks that during near-term pregnancy, a part of the endometrium becomes inflamed, which releases prostaglandins. This results in action potentials firing in the cells, which penetrate into the uterus via these bridges. They dubbed this effect the "myometrial-placental pacemaker" region.

Of course, this was all in rats, which give birth to litters in which each pup is expelled one by one, rather than a single baby (or sometimes two) in humans. Still, Blanks thinks that this result could extend to humans, as there is enough evidence of similar bridge-like structures that connect to the placenta, but more work is needed to be certain. The exciting part of this research is that it all points toward the growing body of evidence that the implantation of the embryo is such a critical moment not only for a successful pregnancy but also perhaps for a successful delivery too.

A full understanding of what is happening in the uterus will need further investigation in both uterine tissue lab experimentation and theoretical modeling. Only then could we have an answer for not only how, why, and where contractions begin but also begin to develop potential strategies to deal with problems such as preterm birth or labor dystocia in which contractions begin but fail to establish. And, what all these experiments and models point to is how much closer we are to unveiling the mysteries of the uterus. "The uterus is like the alien within," Blanks said, "but the picture we have of it is becoming much less hazy."

LABOR DAY

One day in the early 1960s, the US engineer George Blonsky and his wife, Charlotte, were visiting one of their favorite places: the Bronx Zoo in New York. As the pair wandered around the familiar attractions at what is still one of America's largest zoos today, they made their way to the elephant enclosure, where they noticed an elephant strangely twirling around in a circle. The animal seemed in distress, and, concerned about its welfare, the Blonskys approached a zookeeper to ask what was going on. He reassured them that everything was fine and noted that the elephant was pregnant, and what they were seeing is generally what happens when elephants are about to give birth.* The Blonskys watched bemused as this birthing dance unfolded and kept thinking about it as they wandered around the zoo later in the day.

As they left to go home, the elephant's display sparked a eureka moment for George, and after discussing the idea with his wife, the pair began to think whether a similar movement might help humans. After all, childbirth can be slow and painful for women, so if something could be developed to speed things

---

* It is not actually true that elephants go around in a circle to give birth; maybe the zookeeper was trying to get them to leave him alone.

up, it would be revolutionary (pun intended). The Blonskys presumed that the elephant was trying to use the laws of physics to propel the 100-kilogram (220-pound) fetus out of its birth tract. As the elephant carried out the dizzying twirl, the calf would be subjected to a "centripetal" force—the same inward force you feel when you sit on the edge of a merry-go-round at the park (more about the physics of toys and playgrounds toward the end of the book).*

The Blonskys got to work designing a machine that would apply this same force to a woman in the later stages of birth. They soon came up with a basic blueprint that involved a pregnant woman sitting in the "birth position"—legs apart, hands on knees—and strapped to a circular table that was then rotated around in the circle, just like a merry-go-round. The centripetal force, together with the contractions, would extract the newborn, flinging it off on a tangential trajectory to be caught in a "pocket-shaped reception net." The net even had a bell on it that rang once the baby landed—just in case no one was paying attention.

The Blonskys calculated that for the spinning to have the desired effect on the baby, the pregnant women would need to bear large forces. When astronauts blast off from Earth, they experience a "*g*-force" of about 3 *g*, which is the equivalent to three times the force of gravity we are normally exposed to on Earth.† Part of astronaut training programs involves being hurled around in a circle by a centrifuge—similar in design to the Blonsky birthing table—to simulate the *g*-forces that astronauts will encounter on blastoff and beyond. Yet, the Blonskys' device could operate at a maximum of sixty rotations per minute, resulting in a top *g*-force of 7 *g*, double what astronauts experience and what would certainly result in blackouts if ever experienced. It's hard to imagine that it would be much fun to be flung around in a circle at that speed

---

* Strange to think that the force when moving in a circle is pointed inward—centripetal. The centrifugal force—one that is pointing outward—is known as a "fake" force, i.e., it is the same force as the centripetal force; it just depends on which "reference frame" you use. As an observer, the force is centripetal, but if you yourself are doing the rotating, then you "feel" it as an outward force.

† 1 *g* is equal to the conventional acceleration due to gravity on Earth and is given as 9.81 meters per second squared, or ms^2.

while trying to give birth. Despite this, on November 9, 1965, the Blonskys were granted US Patent 3216423 for an "apparatus for facilitating the birth of a child by centrifugal force."[1]

The patent contained ten detailed pages on how to build the machine, which comprises 125 components, such as a concrete floor slab, a lot of wing nuts, a variable-speed vertical gear motor, a butt plate, and pillow clamps.* In the patent, the Blonskys suggested that their device would be especially suited for "civilized women" who "do not have the opportunity to develop the muscles needed for childbirth" while "more primitive peoples" would already have had "ample physical exertion all through the pregnancy," providing them with "all the necessary equipment and power to have a normal and quick delivery"(!). Thankfully, no one has ever built the machine as the Blonskys intended. Although, in 2014, the Science Gallery in Dublin built a full-scale replica of the Blonsky apparatus as part of its "Fail Better" exhibition, which showcased thought-provoking ideas that failed spectacularly.[2]

With genuine health care advances bringing us to the twenty-first century, childbirth is safer now than it has been at any point in history—at least, that is, in developed countries. According to data from the World Health Organization (WHO), the number of maternal deaths per hundred thousand live births in the United Kingdom in 2017 stood at seven, while in the United States, it was nineteen.[3] Yet, it remains stubbornly high in developing countries where it can be as much as five hundred deaths per hundred thousand live births, putting the average maternal death rate worldwide at 211 per hundred thousand live births. Still, the rate in developing countries overall is a 38 percent drop compared to the rate in 2000. The lower numbers in developed countries are thanks, in part, to education, as well as to medical advances, such as caesarean section (C-section), used to deliver a baby through incisions

---

* George and Charlotte Blonsky were posthumously awarded the 1999 Ig Nobel Prize in the field of managed health care for the birth machine, and it even inspired an opera.

in the abdomen and uterus.* Around 20 percent of babies around the world are now delivered by C-section, twice as many as in 2000 (12 percent).[4] In some countries, it is much higher, at over 50 percent: 58 percent in the Dominican Republic, 56 percent in Brazil, and 52 percent in Egypt.[5] Work in 2017 that examined twenty-one separate studies of C-sections worldwide found that they were more likely to be performed on women who were privately insured.[6] Some of that, rather cynically, is because hospitals can charge insurers more for C-section deliveries, but it is also due to risk aversion when labor slows or does not go quite as expected. Yet increases in C-sections do not necessarily translate into healthier babies.[7] Of course, there are times when a women chooses to have a C-section or the procedure becomes medically necessary for the safety of both mother and child. All else being equal, however, babies born from C-sections may be missing out on certain benefits, especially in the first year of birth. Newborn guts are basically blank slates, but as the baby passes through the vagina, it is slathered in microbes, which greatly influences the baby's physiology and lowers the risk of disease. A study in 2010 found that C-section babies received their microbes not from the fauna on offer in the vagina, but from the mother's skin and the hospital environment.[8] Although babies born from C-sections eventually develop a "normal" microbiome consisting of trillions of beneficial microorganisms that live in and on our bodies, it is not clear what this earlier difference—which can last for the first three years of life—means longer term.

It is generally thought that the starter microbes from the vagina and birth canal play an important role in the profile of future communities of gut fauna, which could, perhaps, explain why C-section babies are more likely to develop health conditions, such as allergies, asthma, and obesity later in life. By

---

* You might think C-sections are a recent medical intervention, but the technique has evolved over hundreds of years. It has been mentioned in many ancient cultures, but mostly as a procedure to save the baby rather than the mother. The first time it was documented where the mother survived was in 1826 in South Africa. The procedure was carried out by the British surgeon James Barry and—wait for it—it didn't include any anesthetic. Instead, women were supplied banana wine in presumably copious amounts.

following seven hundred children through the first year of life, researchers in 2020 found that children born via C-section had double the risk of developing asthma and allergies than those born vaginally.[9] Scientists have tested ways to lessen the first-year effect for C-section babies. In 2016, scientists "seeded" C-section infants with their mothers' vaginal microbiota. The researchers placed a gauze inside the mother's vagina an hour before the C-section operation and then just before surgery placed it in a sterile container. Within three minutes of the baby emerging following delivery, the doctors spent a minute rubbing the gauze all over the newborn's body, including its lips and face. Whether it helped to seed a microbiota like vaginally born babies are exposed to, however, was inconclusive.[10]

Yet, as anyone who has been present at birth will know, delivery often results in pushing out more than the baby, just in case you were wondering what the fishing net next to the birthing pool was for. Research in 2020 showed that there could be a fecal element to seeding the microbiota. In their "do-not-try-this-at-home" experiment, the scientists obtained fecal samples from mothers before they underwent a C-section and then mixed the live fecal bacteria, consisting of about a million cells, with milk. When the babies were born, they were then given this rather unappealing concoction for their first milk feed. By analyzing the newborn's excrement three weeks later, they found that the babies had a gut microbiota that resembled those delivered vaginally.[11] The researchers suspect that, during birth, given how close the anal canal is to the birth canal, the babies ingest small amounts of fecal matter that contributes to the seeding of diverse microbiota.

There may be certain benefits for a baby of a vaginal delivery, but the same cannot always be said for women, who can sustain lifelong injuries. The most common is the perineal tear that impacts the perineal muscle, which is located between the vaginal opening and the anus. Such tears come in various degrees, from level one to four depending on the damage. A first-degree tear is small and skin deep that usually heals naturally, while a fourth-degree tear reaches into the anal canal itself. Most first-time mothers, around 90 percent, will tear to some extent, and my wife was no different, but only around

3 percent will have a third- or fourth-degree tear.[12] Risk factors for tearing include firstborns, having a large baby (over 4 kilograms, or 8.8 pounds), induced labor, and whether an instrument is used such as forceps. Delivery by forceps is four times as likely to result in damage to the pelvic floor muscles—a series of muscles that span the area underneath the pelvis.

To ease the baby's journey through and hopefully stop any further tearing, doctors sometimes make an incision, called an episiotomy, to open the vagina a bit wider. Thankfully, most tears or cuts can generally be sutured and heal within a couple of weeks. But the impact of childbirth on the stretching of the inner muscles can take much longer to heal, if at all. Around 20 percent of women suffer urinary incontinence following a vaginal delivery[13] while around 3 percent are estimated to suffer fecal incontinence.[14] The latter can be caused by third- or fourth-degree perineal tears as well as damage to the anal sphincter, a set of muscles at the end of the rectum that surrounds the anus and controls the release of the stool.

Urinary incontinence, on the other hand, is mostly due to damage to the pelvic floor. This often results in leaking urine following sneezing, coughing, laughing, or after strenuous exercise such as carrying a heavy buggy to the car. Complete bladder control can take between three to six months and sometimes longer. But damage to the pelvic floor muscles can also initially go undetected, only to emerge in later life via pelvic organ prolapse, in which one or more of the organs in the pelvis, such as the uterus, slip down from their normal position and bulge into the vagina. Women with pelvic floor damage are around twice as likely to suffer from pelvic organ prolapse.[15]

While the creation of a centripetal birth machine would likely make such injuries a whole lot worse, a better understanding of the mechanics of childbirth may help to not only reduce the need for surgical intervention and "pull-harder" methods of delivery, such as the use of forceps, but at the very least help reduce overall injuries. Could there be a more scientific, and frankly more intelligent, way to manage childbirth?

Once the contractions have fully dilated the cervix to 10 centimeters (almost 4 inches), there is what's known as the second stage of labor, which is more like an Usain Bolt–style hundred-meter race. This is the "push" part: the mother is told to raise her intra-abdominal pressure by bearing down with her muscles and timing her breathing to coincide with the uterus contracting. Together, these movements increase the expulsive force applied to the baby to help push it out. For first-timers, the second stage lasts for around an hour. For those who have been there and done it before, it is likely to be quicker, even as rapid as nineteen minutes[16]—giving a good chance that when the action happens, the birth partner is in the restroom or has gone to the car to fetch the diapers.

Despite the relatively short length of the second stage, it is a crucial time for both baby and mother. This is when the baby begins to descend the birth canal, carrying out several maneuvers, called cardinal movements, to get through the pelvis (Latin for "basin"). As the baby occupies the maximum space in the uterus, the head is initially flexed down so that the chin touches the chest. But as the baby goes through the pelvis, the head needs to rotate to fit through before the neck extends to further push the rest of the newborn's body out of the vagina. Once the head is out, the baby makes a final rotation of the body to help the shoulders fit through the pelvis. This moment is crucial given the risk of shoulder dystocia, which affects around one in two hundred births[17] and occurs when the infant's shoulders become lodged behind the mother's pelvic bone, often because the baby is proportionately too big for the birth canal.

All of these movements are necessary due to an evolutionary arms race between having a pelvis wide enough to allow big-brained babies to safely pass yet narrow enough to enable women to walk efficiently. According to evolutionary biology, when humans evolved to walk on two feet, our pelvises underwent a radical change and became much smaller and compact so that we could better balance on two feet (as discussed in detail in chapter five). Bigger brains came later, creating what is known as the "obstetrical dilemma"—big brains pushing for wider birth canals while the ability to walk on two feet coerces narrower canals. To help deliver all these big-brained babies, the inlet of the female pelvis, the large aperture seen as one looks down at the pelvis, is generally wider and more open than the male pelvis. But even among women, this shape varies considerably.

In the 1930s, two US doctors came up with four general types of the pelvis that were based on differing inlet shapes: gynecoid, android, anthropoid, and platypelloid. They discovered that for "white women," the most common type of pelvis is gynecoid, with about 40 percent of women having this type. It has a pelvic inlet that is wider across than it is front to back, being mostly rounded, and is thought to be the most favorable pelvis type for a vaginal delivery. The second most common type, at 30 percent, is the android, that has a heart-shaped inlet and which is more likely to result in complications and require a C-section. About 20 percent of women have the anthropoid shape, with an egg-shape inlet that is more akin to a male pelvis and can result in longer labor. The least common type is the platypelloid, which is shaped like an American football on its side, being wide but shallow. This type makes a vaginal delivery difficult and often requires a C-section.[18]

It is all very well knowing what the inlet is like—and some dispute such a simple classification as above—but the baby must make its way through the whole pelvis. This pathway is called the "curve of Carus," an imaginary line that goes from the pelvic inlet through the pelvic cavity and out of the pelvic outlet. A UK-led study in 2018 showed that, rather than being anatomically identical, this pathway can vary considerably among women. By measuring the pelvises of 348 female human skeletons from 24 different parts of the world, the researchers found that European women had particularly twisted birth canals, with the pelvic inlet being more oval compared to the round entry for women from sub-Saharan Africa and some parts of Asia.[19] For those women with this twisted geometry, the baby must rotate more in the birth canal to pass through the pelvis. In the 1930s and 1940s, it was reported that African American women had forceps used unnecessarily during birth because the baby failed to turn as expected and doctors thought they needed to intervene.[20] It turned out, however, that these women just had different

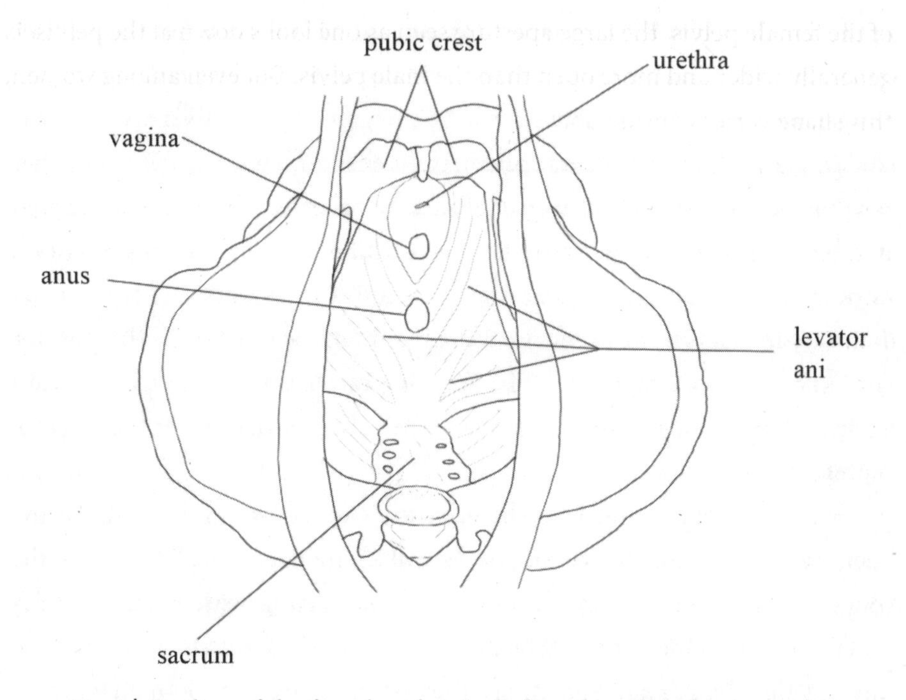

Fig 1 | *Outline of the female pelvis including the pelvic floor muscles.*

pelvic structures, and the babies didn't need to rotate as much as they did for some Caucasian women.

This acrobatic descent through the birth canal has fascinated Jenny Kruger ever since she began studying for a nursing and midwifery degree at the University of the Witwatersrand, Johannesburg. After graduating in 1984, Kruger spent fifteen years as a midwife in South Africa and New Zealand, where she witnessed every possible aspect of childbirth. She was also a keen runner, completing several marathons and half marathons. One aspect of particular interest to her was the struggle that many top-level athletes had during the second stage of labor. Despite thinking that these mothers would be perfect specimens to give birth, Kruger observed that the baby often rotated its head when starting to emerge into the cold light of day, only to stop and not descend any farther. At this point, intervention, such as the use of forceps, was often needed.

It was then that Kruger became fascinated with the mechanics of childbirth, and her thus-far typical career was about to become rather atypical. Kruger went back to school for a master's degree at the University of Auckland in New Zealand to study childbirth in elite athletes. She completed her PhD in 2008 at the same university, focusing on pelvic floor muscles in elite athletes.*. Kruger is now based at the Auckland Bioengineering Institute working on combining engineering techniques to understand the impact that the baby has on the pelvic floor, the main muscle of which is the "levator ani," known more simply as the LA (see diagram). This is a set of three thin muscles that are located on either side of the pelvis (with gaps for the urethra, vagina, and rectum).

The LA is like a funnel for the baby to extend down and goes from the pubic bone in the front to the coccyx, a small triangular bone at the base of the spinal column otherwise known as the tailbone. During birth, the LA muscles initially resist the descent of the fetal head but then stretch to allow passage through the birth canal. Sometimes these stresses are just too high, causing muscle damage, particularly at the pubic bone. There is more chance of damage happening as the length of the second stage of labor increases, which can result in a complete or partial detachment of the muscles from the pubic bone, leading to urinary incontinence and pelvic organ prolapse later in life.

To simulate what happens when the infant's head passes through the pelvis and pelvic floor, Kruger and colleagues took MRI scans of twenty-seven women, which involved over one hundred images for each person. The team painstakingly went through each one to pick out the various muscles of the pelvic floor, including the LA. With this detail, they then created a model of the pelvis, including the bones and muscles, using a finite-element method (explained in chapter five to understand fetal kicks in the uterus). Kruger and doctoral student Xiani Yan also created a finite-element model of the fetal head from computerized tomography (CT) scans of infants that were about

* Why elite athletes have a harder time giving birth is still not fully understood, but some think it is because they have tougher pelvic floor muscles.

nine days old.* While the bony head and pelvis were fixed in place and not allowed to deform,† the muscles of the pelvic floor could resist the head and then stretch to accommodate it.

Using muscle stiffness values from experimental data, they ran their simulation to virtually "deliver the baby," with the fetal head making its way down through the pelvis.[21] Previous models had often prescribed the path that the head takes—i.e., the curve of Carus, as it passes through the birth canal. But Kruger and colleagues did not put any such constraints on the head as it passed through. When they let the simulation progress, they found, incredibly, that the infant head engaged with the pelvic floor, but then as it went through, the head turned on its own accord, simulating exactly what happens during birth. "It was really exciting to see the head turn in the model. Somehow the baby finds the 'energy minima' as it descends through the canal—it's physics!" Kruger explained to me.

Running their calculations, Kruger and colleagues predicted that the force required to get the head through the birth canal increases as the head passes through, reaching a maximum of around 30 N[22].‡ By including and removing various muscles and rerunning the simulations, they could determine what the most important factors are in lowering the force needed for delivery. The biggest factor, as you might expect, is the size of the fetal skull, but surprising features emerged as well, such as the force increasing for a smaller area of the levator hiatus, or the "gaps" in the LA muscles for the vagina or anus, that can stretch around 2.5 times during birth. Smaller gaps mean more force than larger gaps.

---

* A CT scan combines a series of X-ray images taken from different angles and uses computer processing to create cross-sectional images of bones, blood vessels, and soft tissues, offering more detailed information than X-rays do.

† This is not strictly true for the infant head. We know that it undergoes huge deformation during labor, sometimes appearing like a "sugarloaf."

‡ As we learned in chapter five, this force is about the same as that a two-year-old child can exert when pushing with a thumb.

One aim in this preliminary research is to produce a way to quantify the risk of injuries that take place during birth. This could be done via personalized risk assessment that would be based on a small number of characteristics, such as the size of the fetal head, the outline of the bony pelvis, and an estimate of LA muscle stiffness. The first two would be possible to obtain from routine ultrasound scans, but Kruger admits that getting a measure of muscle characteristics would be tricky given the difficulties of carrying out such examinations on pregnant women. If this were possible, however, then the model could be used to calculate a personalized force map for delivery. Kruger realizes that performing a finite-element calculation would take a huge amount of time—as much as a week for one person, which is clearly unfeasible in a clinical setting. To address this point, her team has already come up with a statistical technique that would take just minutes instead. The method uses a small number of inputs for the geometry of the pelvis and head, but then calculates birth forces based on statistically extrapolating those already carried out with finite-element calculations.[23]

There are still many issues to be resolved with the modeling, such as getting a better understanding of the mechanical properties of the pelvic floor muscles, all of which would result in a higher estimated calculated force. But it may, with improvements, provide a method to help clinicians create better birth plans. "Whenever we tell clinicians about these models, they love it," Kruger said. "But we first need to be confident that we can give them reliable predicators before it can be used in a clinical setting."

As anyone who has witnessed childbirth knows, labor is not just a case of infants passing through solid muscles and bone. There are a lot of fluids at play, such as blood and amniotic fluid. Mechanical engineer Megan Leftwich thinks that the effect of fluids is a hugely overlooked aspect of childbirth. Much of Leftwich's previous and current research is on the mechanics of swimming animals. For her PhD at Princeton University, she studied snakelike sea lampreys, creatures straight out of a horror film with circular mouths filled with razor-sharp teeth arranged in consecutive circular rows. She discovered that the flexible bodies of these animals allow them to power through water, using their fins to do so. Much of her work today involves the swimming ability of seals, which propel themselves via the front flippers as opposed to anything at the back of their bodies, such as a tail.

During her studies of the vampire-like sea lamprey, Leftwich became pregnant and, given that her PhD centered around fluid dynamics, she became curious about the mechanics of what she was about to go through: childbirth. She began to search the scientific literature for research on the fluid mechanics of birth, but to her surprise, found extraordinarily little. In the meantime, her son was born, and once she had come to grips with diapers, feeding, and not sleeping at all, she was back in the lab, completing her doctoral degree in 2010. Two years later, she moved to the George Washington University in Washington, DC, where she began to think about ways to study birth from a fluid dynamics perspective.

Leftwich had an inkling that fluids in the uterus could be a critical factor in lowering the force necessary to deliver a baby but needed a way to test it experimentally. The main fluid at birth is the amniotic fluid, which occupies about 20 percent of the uterus. It is mostly made up of water, around 98 percent, with the rest consisting of mostly lipids (fatty-acid compounds), proteins, and glucose. At large volumes like in the womb, the behavior of the amniotic fluid should be like that of water. But at smaller scales, such as when compressed by the baby's head in the birth canal, it can take on strange characteristics, thanks to the 2 percent of fats and other material that allow it to play a big part in easing the baby along.

To understand the effect that this may have, we need to examine the behavior of liquids under the effect of a force, which was first explained in the late 1600s by Isaac Newton. As well as formulating laws of motion and gravitation, he was the first to use mathematics to model the movement of fluids, such as what happens to a liquid when you put some stress on it by changing how fast it flows. Newton discovered that different liquids behave very differently. For some fluids, the viscosity does not change no matter how much force is applied to them. These are called Newtonian fluids, and examples include water and air. In this case, the viscosity stays roughly the same whether water is trickling down a windowpane or flowing fast down a river where it is subjected to increased forces from the flow.

Fluids not considered Newtonian are called non-Newtonian fluids, which change their viscosity depending on the amount of force applied. If you are bored at home one day or find yourself locked down again because of a pandemic, you can make the perfect non-Newtonian fluid. All you need is to mix one part of water and 1.5–2 parts of corn flour, and the result is a weird blob of fluid that acts very strangely when you deform it. It is guaranteed to keep a toddler entertained for at least five minutes, but it will certainly take a lot longer to clean up. Non-Newtonian fluids come in many types depending on how they respond to a force applied to them. Some become "thicker" under stress while others become "thinner," or flow faster. The latter is known as "shear thinning." This means that the viscosity of the fluid decreases when it is deformed, or when a shear force is applied.

Shear-thinning materials, such as paint, tend to be jellylike. Paint can be quite thick when you put it on your paintbrush, but when you start to put it on the wall (by applying a force on it), it thins, making it much easier to apply. Another example is ketchup. If you hang a bottle of ketchup upside down, some may come out, eventually. However, if you vigorously shake the bottle—or, better still, hit the neck of the bottle—you will manage to get more out. As a force is applied to the ketchup, the ketchup thins out slightly, its viscosity decreasing, making the condiment easier to get out of the bottle. Such shear-thinning non-Newtonian fluids are also found in nature. In 2017, for example, scientists found that frogs capture prey using shear-thinning saliva. When the saliva hits the insect, it thins and spreads over the insect so that it is stuck in the clutches of the frog's retracting tongue.[24]

To investigate this possible viscous effect on the mechanics of birth, Leftwich built an artificial latex uterus, about a quarter the size of a real uterus at full term, that was like a balloon with a hole at the bottom. Her team suspended the "uterus" in the center of a frame that was held in place by unstretchable strings and filled with water to match the 18 percent volume of the amniotic fluid. She then inserted a solid oval piece of wood to represent the baby's head in the uterus, keeping the same size perspective of a quarter of the average newborn head size. The piece of wood had a small hook on one side so it could be pulled out of the uterus with the force measured by a standard force gauge.[25]

Leftwich and colleagues loaded the head at different angles, with zero degrees corresponding to what is thought the ideal scenario, in which the fetal and maternal spines are aligned. This took the least amount of force to extract from the uterus, which is expected given that it is the narrowest part of the oval going through the hole. But when the angle of the head was increased to around 20 degrees—shifting away from this perfect alignment and increasing the width of the baby's head as it engages with the hole of the uterus—then, as you might expect, the force slightly increased. This amount of misalignment is reported to happen more than you might think. A study in 1993, for instance, found that around 25 percent of births have a head misalignment of around 20 degrees.[26] So far, so easy (ish). But when the angle went beyond this—around 30 degrees or so—then the force required to get the head out increased rapidly, showing that at such an angle it would never be possible to extract a baby using methods like forceps, which often involves a brute force, "pull-harder" approach.

But that was using water, so Leftwich and colleagues began to investigate what happens when the viscosity of the fluid is increased, doing so by replacing water with white vinegar. Leftwich and her team thought that increasing the viscosity would increase the force needed to pull the "baby" out. Yet, they found the opposite. Increasing the viscosity of the amniotic fluid by 30 percent decreased the pull-out force by an average of around 50 percent. This was a sizable difference, and Leftwich thinks that this possible non-Newtonian effect could be at play during birth. The baby squeezes the amniotic fluid, changing the fluid properties and, therefore, decreasing the force needed to push out.

Of course, the amniotic fluid is not the only fluid present at birth. Yes, there is a lot of blood (which is also non-Newtonian), but there is also a strange wax- or cheese-like white substance called "vernix caseosa" that coats the skin of newborn babies. This is what midwives rub off with a towel—along with other stuff—once a newborn is born. The vernix has several roles when in the womb, notably protecting the unborn baby's skin from infection as well as from the amniotic fluid, as without this protection the skin would wrinkle and chap. The vernix caseosa contains a greater volume of lipids and proteins than the amniotic fluid, and Leftwich suspects it is highly viscous and non-Newtonian in fluid form. Given that viscosity plays an important role in birth, Leftwich suspects that this cheesy substance could act like a lubricant during birth and perhaps be more important in delivery than the amniotic fluid.

Leftwich admits her setup is relatively simple but this is because, from a fluid-dynamics perspective, we know so little about the role that fluids play during birth. Leftwich compares the situation to what it was like in the 1970s working on the fluid mechanics of the heart. Only with decades of research could ideas be refined and improved to arrive at the incredibly advanced models that we have today. "A lot of research initially on the uterus was based on getting immediate outcomes, or reducing the instances of maternal death, rather than on the process itself and how that could be improved," Leftwich said.

In that sense, the experimental models of birth need to be simple to get an idea of the important features before they can become more complex. Perhaps one of the biggest aspects in improvement strategies is to make birth models more realistic—be they Kruger's pelvic floor simulations or Leftwich's fluid dynamics experiments—to include the huge deformation that occurs to the infant's head as it passes through the birth canal. A baby's skull is not a single bony plate like an adult's, but is instead composed of several sections joined together with tough tissues called sutures.* These sutures allow the plates of

* When several of these sutures meet, they create a fontanel, which plays a key part in making the skull flexible enough for the brain to grow. A newborn will have several fontanels on the skull, but the ones on the back of the head and the top are the most well known. It's fairly easy to feel the fontanel. If you run your finger from the center of the

the skull to slide over each other, in a similar manner to how tectonic plates in the earth move due to conventional forces in the crust.

The head is so big that it literally gets crushed as it exits. When the baby passes through the birth canal, the head can change shape by as much as 10 millimeters (slightly less than half an inch), or 10 percent of the average-sized skull, to allow for an easier passage through the birth canal.[27] When our second child was born, he had a particularly "sugarloaf" head (despite the deformation, the worst of it usually disappears within a couple of days). Most of the skull snaps (not literally) back into shape soon after birth, but for some, it does not, and they retain the sugarloaf top (sort of like an elongated cone with a rounded top) for some time, although without any noticeable health effects.

There is a lot more to learn, and Leftwich is now planning to increase the complexity of the experiments by using a 3D printed baby's head that has a soft, deformable "skin," like a silicone gel, in the latex uterus. She also plans to eventually use a pelvis-shaped structure as well as a uterus that is filled with actual amniotic fluid—although this would be far from straightforward given not only the difficulty of obtaining the legal consent from hospitals and patients to do so but also the practicalities of collecting it from pregnant women. But those challenges are unlikely to deter Leftwich. "If a better understanding of the fluid mechanics of birth leads to even a few percent of women not undergoing unnecessary C-sections or the use of forceps," she said, "then that would make it all worthwhile."

---

top of the forehead toward the back of the head, you will feel a depression somewhere near the very top of the head.

SECOND INTERLUDE

THE ENGINEERING OF MODERN DIAPERS

Well before the baby is born, you may find yourself shopping for things that, in the end, you don't really need, but one item that will definitely be required at birth is some sort of diaper. Indeed, one of the first acts of parenting you will likely perform is putting a diaper on your newborn. And, chances are, it may be the first time you have ever done so, just to add to the nerves as everyone looks on.

When the time comes, you'll likely grab a diaper from the hospital bag (if you are using disposables), lie the baby on its back, lift both little legs up, and slide half the unfolded diaper underneath the baby. Then lower the legs, bring up the front of the diaper to the stomach, and attach the sticky tabs at the sides to fix it all in place. Easy. Then realize the nappy is on back-to-front and do it all again.

The first content in the diaper, along with some urine, will likely be meconium, a tar-like substance that makes up babies' first poop. It is sticky, thick, usually dark green—and thankfully odorless. It consists of materials that have been ingested during the baby's time in the uterus, such as cells, mucus, amniotic fluid, and bile, as well as lanugo, the fine, small hairs on the fetus's skin. If there is (only) one piece of advice in this book, let it be this: before you put the diaper on, apply a thin coat of petroleum jelly on your newborn's dry bottom, which will make it much easier to clean the meconium off. While most new parents are braced for meconium, it can still be surprising how long it takes to clear out of the baby's system, so keep up with the jelly.

Swaddling babies for comfort but also for protection from their own urine and feces has been carried out since ancient times. This involved taking strips of material, such as cotton or linen, and winding them around the body, with the strips sometimes being left in place for several days.[1] Another option, especially in warmer climates, was to not bother with cloth diapers at all. Parents would wait for the infant to commence a bowel movement while feeding and then hold the baby over a container or in a bush. This latter technique avoids one major problem with cloth that is exposed to urine and feces—it sticks to the delicate newborn skin, causing conditions such as dermatitis, the itchy, blistered, and cracked skin that still affected up to 25 percent of diapered babies in the early twentieth century.

There were many innovations that led to the diaper that we all know today, starting off with the invention of the safety pin in the mid-nineteenth century that prevented leakages by safely providing a tight fit around the newborn. The 1940s led to several more advances, starting with the creation of the disposable diaper. It started out rectangular in shape and had a plastic covering made of a nonwoven material.* Inside were layers of paper tissues that absorbed the urine. This layering system allowed around a hundred milliliters (about 3.4 fluid ounces) to be absorbed before the diaper leaked and had to be changed.[2] The 1960s and 1970s brought the introduction of a cellulose pulp core, Velcro fasteners, and an hourglass-shape diaper, as well as using thermoplastic polypropylene, which is more comfortable for the baby's skin. In 1979, the US consumer goods giant Procter & Gamble—owner of the Pampers brand—filed a patent for a wetness guide based on a pH indicator made from

* This is a fabric that is not knitted or woven and usually made by bonding fibers.

bromophenol blue. This dye is yellow when dry, but when it encounters the alkaline pH of urine, it changes to blue, which at least gave some indication of when the diaper had reached capacity. An innovation in 1987 introduced elasticized side pockets to the diaper, which improved its ability to contain feces, even "explosive liquefied bowel movements."

Despite those improvements, however, diapers are still not totally unassailable. As every parent knows, there will come a time when the baby produces another level of fecal explosiveness that is beyond even a modern diaper's capabilities. When this occurred, and it was more often than I would have liked, my wife or I would shout "BREACH!" and the other person dropped everything, including another child, to come and help. Often the "breach" resulted in liquid feces escaping right up the baby's back (usually, because the baby was mostly lying down) to fully soil whatever clothes the child had on, which were usually that extra-special dungaree set bought by grandma. This happens because diapers are not really designed to catch large volumes of diarrhea that escape up the back. Instead, what they really excel at is soaking up huge quantities of urine, all thanks to some clever chemistry and engineering.

Diapers are big business. The global disposable diaper market is estimated to be worth about $65 billion by 2025.[3] If a baby wears disposable diapers exclusively, expect to buy at least 2,500 diapers before the end of the first year. Before finally being potty trained (which is another challenge altogether and well outside the scope of this book), each child will require thousands more.[4] Disposable diapers result in thirty-nine million tons of waste worldwide,[5] and in the UK, for example, they are responsible for 2–3 percent of all household waste.[6] In a house with a child in diapers, disposables make up a whopping 50 percent of household waste, and they are the third largest single consumer item in landfills.[7] Given they take about five hundred years to decompose, you would rightly think they are bad for the environment. That has led some to use reusable diapers, which can be washed and reused rather than just thrown away after a single use.

However, a study in 2008 by the UK's Environment Agency found that the carbon emissions impact of reusable diapers is not as good as you might think, mostly because of the energy requirements to launder so many reusable diapers.[8] In his book *How Bad Are Bananas?*, Mike Berners-Lee reaches a similar conclusion.[9] He found that a disposable diaper creates 130 grams (4.6 ounces) of carbon dioxide equivalent,* but washing a reusable diaper at 90 degrees Celsius (194 degrees Fahrenheit) and then tumble drying produces 165 grams (5.8 ounces) of carbon dioxide equivalent.† The best option is to wash the reusables at 60 degrees Celsius (140 degrees Fahrenheit) in a large load, dry them on a washing line, and then pass them onto a second child. This process produces the equivalent of 60 grams (2.1 ounces), or less than half the amount, of carbon dioxide as a disposable diaper. In the end, there is no magic diaper when it comes to which type to use, and to put it all into perspective, one family holiday by plane would undo the carbon savings of perfect nappy practice many times over.

The reduction in energy demand for disposable diapers is mostly due to innovations that have made them more compact and thinner than previous incarnations. The effectiveness of today's diapers is thanks to their multilayered structure that guides the urine away from the baby's skin. This is down to three layers in particular. The thin top layer, or the layer that is closest to the baby's skin, is made from polypropylene, which acts to pass the urine down into further layers.

This mesh-like layer also repels water, and while this might seem strange—after all, the whole purpose of the nappy is to absorb water—the clever trick is that this aspect depends on how fast the liquid is moving. When a liquid is squirted onto it—as happens when a baby urinates at around two meters per second, or five miles an hour—the urine passes through. Yet, if a droplet

---

* This is a measure of the total climate change impact of all the greenhouse gases caused by an item or activity in terms of the amount of carbon dioxide that would have the same impact.

† This, of course, depends on the energy mix of the country in question. The figures used here are for the United Kingdom's energy mix.

is gently placed on the surface, it does not get absorbed and stays on top. The squirted urine passes through while the liquid that is held inside the diaper does not pass back through at all, keeping the baby dry.

The next section, heading deeper inside the diaper, is the so-called surge layer, also known as the acquisition layer. This is another engineering marvel, as it contains three different sheets, each one having a different-sized hole for the liquid to pass through. This all has to do with capillary forces that we will learn more about in chapter nine. In the uppermost sheet, the size of the holes is relatively large compared to the middle and lower ones, which have many tiny holes.

This arrangement allows the acquisition layer to do three main things. One is to act like a one-way valve—letting liquid pass from bigger to smaller holes but not the other way around. Another is that the big holes let the fluid in quickly, so out of the way of the baby's skin. The third is that—thanks to the capillary force—the smaller holes in the lower sheets help to spread the liquid around the diaper. This means that while urine may enter the diaper at a given point or area, by the time it reaches the bottom of the acquisition layer, it is nicely spread out (in theory, at least) and ready for the third and final layer, which is where all the chemistry happens.

An infant produces about 15 milliliters (or about a half of a fluid ounce) of urine per hour, on average, so nine hours would result in 135 milliliters (about a half cup). For an infant to stay dry throughout the night, the diaper needs to hold a similar amount of urine. If it did not, the infant would wake up fussing about wet legs or a wet bed and that would be no good to anyone, especially sleep-deprived parents having to change the bedsheets again. Diapers, then, need a material that is exceptionally good at absorbing water, or urine.

In the 1960s, researchers at the US Department of Agriculture were searching for such a material, but for retaining moisture in the soil. Early attempts included "grafting" polymers* onto starch molecules that expand

* Polymers are long-chained molecules formed from many small, repeating motifs.

when water is added, turning the starch into a thick, transparent gel. Their efforts resulted in a new polymer called sodium polyacrylate, which turned out to be a so-called superabsorber, able to hold hundreds of times its weight. It took a decade or so to find consumer applications, first in sanitary pads and incontinence products. In 1980, the US multinational consumer firm Johnson & Johnson filled a patent to use the substance in diapers.[10] Other companies also began to use this super-absorbing polymer, including Procter & Gamble in Pampers.

One of the biggest difficulties concerning diapers—besides trying to put one on a kicking (and sometimes screaming) infant—is deciding when they are full and need changing. The pH indicator can be rather crude, and in 2013, students from Attleboro in Massachusetts carried out an experiment to test it. They used salt water to mimic urine, finding that the "capacity" as shown by the indicator varied between 50 milliliters to 85 milliliters (or between about a quarter and a third of a cup)*—lower than the actual maximum capacity.[11] This means that there is possibly a lot of unnecessary waste, with parents changing diapers that are nowhere near full.

To tackle this issue, researchers at Massachusetts Institute of Technology are developing a smart diaper that not only measures when the diaper has reached capacity but also is able to let you know when it's time to change it—which could come in handy when the diaper is hidden away under various onesies. The smart diaper contains a moisture sensor placed below the super-absorbent polymer layer.[12] When the layer gets wet, it expands and becomes slightly conductive, albeit very weakly. The team incorporated a small bow-tie-shaped radio frequency identification tag that emits radio waves, and when this interacts with the water in polymer layer gel, it generates a small current that is picked up by a sensor. This is then used to send a radio signal to a reader, which can be placed up to a meter away. The reader, connected to the internet, then sends a message to your cell phone to notify you that the diaper needs changing.

---

* The Huggies diaper that was tested did actually reach 130 milliliters (about 4.5 ounces) until it showed it was full.

The researchers say that the tag is low-cost and disposable and can be printed on rolls of individual stickers, like barcode tags. Given this, such tags could be incorporated into every diaper, where it could also be used to detect health problems, such as constipation or incontinence. As a first step, these diapers may be well suited for use in neonatal units, where midwives care for many babies. And diapers aren't just for babies, of course. Such tags could be used in adult diapers, for patients who may be too embarrassed to report themselves that a change is needed.

Whatever technological innovations lie around the corner, one aspect that will never be solved by science is whose turn it is to change the diaper. One can only speculate that this issue pushed Iranian engineer Iman Farahbakhsh from Amirkabir University of Technology in Tehran to such an extent that he designed a device that could do it for him. His diaper-changing machine for babies, which was awarded a US patent in 2018,[13] is a washing-machine-sized contraption that involves putting a baby inside on its back on an inclined slope. The baby is constrained via several "safety" belts to "prevent the infant from rolling off." Then a fork-like (!) arm somehow removes the soiled diaper from the baby and puts it into a bin located inside before two sprinklers near the seat wash the baby at the "desired water temperature."* A dryer then finishes off the process before a new diaper is "rolled" around the baby via the arm and a clamp.

According to the website, the contraption costs about $1,000 and is suitable for infants aged between three months to five years. A diaper change takes two minutes. If it ever came to market, it might just be worth the cost alone when dealing with those diarrhea-soaked incidents.

---

* Farahbakhsh bagged the 2019 Ig Nobel engineering prize for his contraption and started a company, BabyWasher: www.babywashers.com.

8 THE TREE OF LIFE

I was looking forward to cutting the umbilical cord with our firstborn, Henry. It felt like a coming-of-age moment and it looked easy—at least, that is, from watching someone do it on TV. Apart from supporting your partner going through birth, it is also the only aspect where the birthing partner can do something useful. The problem is that after witnessing childbirth for the first time, you are still looking at all the blood and gore—as well as your newborn—too much to be aware of what else is going on. Suddenly, you are presented with a pair of scissors so the midwife can get on with cleaning up all the mess.

Cutting the cord is harder than it looks. Not only do the scissors seem to be incredibly blunt, akin to childproof versions, but the umbilical cord is surprisingly rigid. The first time around, I did not quite appreciate how tricky it would be, making a series of yanking attempts before finally getting through. For our second child, Elliott, however, I made sure I was ready. That does not mean I brought my own gardening shears along, but rather I made a much more decisive cut. The reason for the cord's hardiness is that the blood vessels in the umbilical cord—the umbilical vein and two umbilical arteries—are encased in a gelatinous substance called Wharton's jelly that is similar in stiffness to rubber. They also coil around each other like a helix, which makes it tough.

Blood travels to the fetus via the single umbilical vein, and blood from the fetus flows back to the placenta through the two arteries. The umbilical cord is connected at one end to the baby and at the other to the placenta, an organ unique to pregnancy that is a direct link between the mother and fetus. For the past six months or so, the placenta has been providing the fetus with oxygen and nutrients—such as fatty acids, proteins, and vitamins—while, in return, the fetus kindly sends back waste products, such as carbon dioxide, hormones, and urea, a major component of urine. After childbirth, the placenta still resides in the uterus and once the umbilical cord is cut the baby is free. But the placenta is still not quite ready to make its grand entrance. It first must detach from the uterus before being "delivered"—known as the third stage of labor.* This takes around ten minutes following the birth of the baby,[1] or sooner if an injection of oxytocin is given. When Henry was born, I was too preoccupied with staring at this new person to observe the placenta being delivered. For Elliott, however, I made sure that I had a good look as it was picked up and dumped into a white dishwashing bowl. It looked like a purple jellyfish that had just been washed up on the beach. With an elaborate network of thick veins, it was an incredible sight, resembling a flat cake,† certainly not for the queasy.

The placenta did not start off this voluminous, of course. Its journey begins at the start of conception, and it is thought that the placental cells are the first to differentiate when the embryo consists of a solid ball of cells, known as a morula.[2] At the blastocyst stage, around five to six days following fertilization, the ball of cells contains two different types of cells: the inner-cell mass that become the baby, and the outer layer, or trophoblast, which develops into the placenta (as explained in chapter three). The trophoblast invades and remodels itself on the wall of the uterus, or the endometrium, and is reminiscent somewhat of a cancer tumor. From just a collection of

* The first stage of labor is when persistent contractions result in the cervix being fully dilated to 10 centimeters (about four inches). The second stage is when the baby moves through the vagina and is born, and the third stage is when the placenta is delivered.

† This is presumably why placenta means "cake" in Latin. Not to mention that some people eat it, claiming it brings health benefits.

cells in early pregnancy, the placenta quickly begins to form a basic structure. This eventually includes a network of fetal vessels that branch out to form miniature trees—a bit like Japanese bonsais—known as villous trees. These are bathed in maternal blood in the "intervillous space." It sounds a bit gory, and it is. The placenta is like putting fifty connected bonsai trees upside down at the top of a fish tank that is full of blood, thanks to the pumping of several arteries at the bottom.

The first critical job of the placenta is to key into the maternal blood supply, a bit like a thirsty vampire, but how it does this needs to be carefully controlled. The source of the maternal blood in the uterus is a strange kind of "spiral" artery, which as the name suggests, is coiled like a corkscrew. The spiral arteries emerge from the uterine arteries and begin in the myometrium and extend into the endometrium (the innermost layer—see chapter six for more details of the uterine layers).[3] Early in the first trimester, chemical signaling attracts hundreds of special cells called extravillous trophoblasts on the tips of villous trees to go on a grand migration to the endometrium to begin invading the uterine tissue. They surround the entry point of the narrow spiral arteries, breach inside, and then travel down the vessels. Here, the extravillous trophoblasts do two incredible things. The first is that they

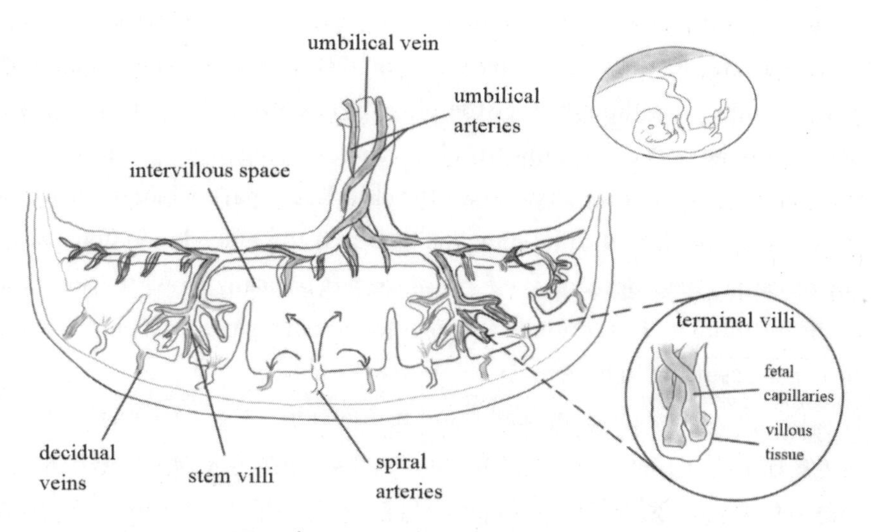

Fig 1 | *Schematic of the placenta.*

widen the diameter of the spiral arteries fourfold, from 50 micrometers pre-pregnancy to around 200 micrometers. The remodeling slows the speed of the resulting blood flow and allows oxygenated blood to slowly percolate throughout, resulting in an incredibly uniform distribution of oxygen in the whole placenta, even to its extremities.

The widening of the arteries is critical because if they remained narrow, then the blood flow would be incredibly fast, creating vortices, or little whirlpools, that would unevenly distribute oxygenated blood around the placenta, eventually constraining the growth of the fetus. A failure to correctly remodel the spiral arteries is thought to be behind preeclampsia, high-blood pressure that affects both maternal and fetal blood-flow rates. Women are checked for preeclampsia during pregnancy. Its impact on the placenta can lead to fetal growth restriction that affects about 5 percent of pregnancies worldwide.[4] Such growth issues for babies can lead to complications further down the line, such as diabetes and even death. And it is not only the growth of the baby that is affected but also the placenta itself. The weight of the placentas measured after the delivery of growth-restricted babies tend to be about 25 percent smaller than those of normal pregnancies.

The second important aspect of this invasion is that the extravillous trophoblast cells travel down the arteries where they clump together to act as a plug to stop the flow of blood. The existence of these plugs, which are about 7 millimeters (just over a quarter of an inch) long, was first identified in the 1960s,[5] and it was thought that the plugs were entirely impermeable, like a cork in a wine bottle, stopping any blood from passing. But it seems that even when fully plugged, they let plasma—the clear liquid part of blood that carries the blood cells—into the placenta. Yet, if there is too much blood flow early in the placenta's development, the force of the jets basically blows the soft tissue of the terminal villi apart and transports too much oxygen-rich blood into the placenta, resulting in hyperoxia, or oxygen toxicity.[6]

The plugged arteries instead result in a flow into the placenta that is a bit like a sprinkler that you might use to water a lawn. It is not exactly known how the arteries get unblocked, which begins as early as week six of gestation, but it is estimated that between thirty to sixty spiral arteries are opened by the end of the first trimester when the maternal oxygenated blood fully enters the placenta. This is around the time that the placenta takes over from the yolk sac to nourish the developing baby,[7] and so the switch represents a critical time for the fetus, which can result in miscarriage if there are issues with the "unplugging" of the arteries.

Alys Clark from the University of Auckland in New Zealand has been working on the physics of this trophoblast invasion and spiral artery plugging for over a decade. With a PhD in mathematics from the University of Adelaide in Australia, she moved to New Zealand in 2008 where she studied the movement of blood in the lungs. In 2010, Clark was approached by a placental biologist who was puzzled by a postmortem of a baby that had bits of the placenta in the lungs. The biologist wanted to know whether having these chunks of placenta would impact lung function. It was during this investigation that Clark began to see the strong links between human lungs and the human placenta.

Using techniques developed throughout her work on the lung, Clark and her team used anatomical measurements from images of dissected pregnant human uteri from women who had died during pregnancy. This showed the structure of the plugs at various stages of gestation, and Clark used this to build a mechanical model of the blood flow through the spiral arteries to understand how this unplugging happens. Clark and colleagues simulated a portion of the plugged artery as a cylindrical tube that is then jammed with smaller spheres that represent the packed cells. The cells are placed in the tube in a configuration that corresponds to similar cell densities seen in the historical images. The model not only took into consideration the forces at play from the maternal blood flow but also the molecular forces of the plug that keep them together and can also split them apart.

Clark found that the cell-cell adhesion is strong enough to keep the plug intact as a cohesive mass. She also discovered two possible mechanisms for how it could break up later in pregnancy. The first is that the pressure of the blood flow in the spiral arteries gradually pushes the whole plug as a single piece into the placenta. A bit like slowly loosening a cork on a champagne bottle and then letting the pressure inside eventually shoot the cork out. The second possibility is that the plug is progressively whittled away via the creation of individual channels that eventually lead to a fully unblocked artery.[8]

Given the challenges of carrying out actual in situ experiments, it is not currently known which mechanism is responsible for this uterine plumbing fix, requiring further investigation. "There is certainly a mechanical element to breaking down the plugs, but other factors are also likely to come into play, such as the chemical role that oxygen has in breaking the plug," Clark said. "What is crucial is the timing of the blood flow—too early and it results in many problems."

Clark and colleagues have also used data from the Centre for Trophoblast Research in Cambridge, England, which contains an unequaled historical archive of anatomical slides of dissected pregnant human uteri—as well as images provided by the University of Bristol, UK, to examine how the uterus evolves in early pregnancy. Early results from that work show that the plugs can hang around a little longer in the spiral arteries than previously

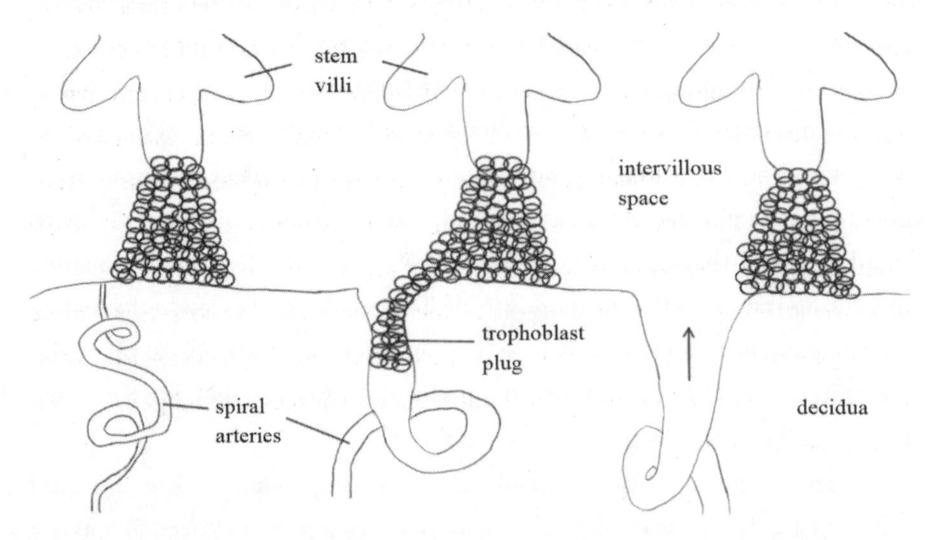

Fig 2 | *Outline of the plugging of the spiral arteries (bottom middle) and the remodeling that subsequently occurs (bottom right).*

thought—up to eighteen weeks. Although, beyond the first trimester, the plugs still allow the blood to flow into the placenta.[9]

Once fully unblocked, however, the arteries pump maternal blood into the placenta at up to 10 centimeters (about 4 inches) per second. Blood flow in these arteries in a nonpregnant uterus is about 40 milliliters (1.4 fluid ounces) per minute, but at full term, the volume of blood flow is eighteen times larger at 750 milliliters (just over 25 fluid ounces) per minute.[10] Given how critical the first ten to twelve weeks are in the formation of the placenta, Clark thinks more focus should be given to it during routine ultrasound measurements.* Currently, such examinations are geared toward spotting where the organ resides on the uterine wall as well as the blood flow to and from the baby via Doppler sonography of the umbilical cord and the largest uterine artery. Clark said that this should be expanded to look at what is happening to the blood supply through the uterus and into the placenta, which could shine a light on potential problems before they emerge later in pregnancy.

The placenta never stops growing, starting off a few millimeters (less than a tenth of an inch) in size. A healthy placenta at full term is around 22 centimeters (8.7 inches) in diameter, 2.5 centimeters (about an inch) thick, and has a mass of about 650 grams (just under 1.5 pounds). The placenta is estimated to contain around 550 kilometers (almost 342 miles) of fetal blood vessels—the same length as the Grand Canyon—giving a surface area for gas exchange of 13 square meters, the same as the floor area of an average-sized bedroom. Part of the difficulty studying the placenta is due to these varying scales, and how this huge network of the fetal vessels, which are each about 200 micrometers across, ultimately affects the performance of a centimeter-sized organ.

* If you do use a fetal Doppler to detect the heartbeat of the fetus in later pregnancy, you will likely pick up a "whooshing" sound, which comes from the placenta.

Modern imaging techniques have shown just how intricate and varied the fetal blood vessels inside the villous trees are. In the past decade, the vessels have been clearly resolved by scanning confocal microscopy, which, in basic terms, uses a microscope and a pinhole to block out-of-focus light in the image and then scans across a sample to be studied. This technique has shown them to be far from perfect cylindrical tubes, instead being incredibly complex shapes, having bulges, kinks, twists, and coils[11]—a bit like what you find in a bag of Cheetos.

Igor Chernyavsky from the University of Manchester in the UK has been studying the placenta since 2008 when he began his PhD studies at the University of Nottingham, also in the UK. He is tackling one of its biggest mysteries: How does this "multiscale" organ manage to efficiently exchange such a large variety of different gases and nutrients both to and from the fetus? The fetal blood vessels in the placenta contain about a quarter of the total fetal blood—about 80 milliliters, or 16 teaspoons, at term. But the maternal blood in the intervillous space and the fetal blood in the vessels do not mix. They are separated by 2–3 micrometers of a thin villous tissue that encases the fetal capillaries.

The exchange of gases between maternal and fetal blood is via diffusion through the villous tissue, with the fetal vessels closest to the villous tissue thought to be doing the exchange. Diffusion is the net movement of things like ions or molecules from a region of higher concentration to a region of lower concentration. In the placenta, gases diffuse due to the partial pressure difference between the maternal and fetal blood.[12] As maternal blood takes up carbon dioxide, for example, this makes the blood acidotic,* which helps oxygen uptake into the fetal capillaries.

Chernyavsky has combined mathematical modeling with the intricate geometry of the fetal blood vessels to understand the transport of gases and other nutrients. His research found that, despite the incredibly complex

---

* Via the so-called Bohr effect, the increase in carbon dioxide results in hemoglobin in the blood releasing oxygen. Given the higher concentration of oxygen in the maternal blood, this then helps to increase oxygen uptake in the fetal capillaries.

topology of the fetal vessels, there is a single dimensionless number—like the Reynolds number explaining the swimming capabilities of sperm—that can explain the transport of different nutrients in the placenta. This is called the Damköhler number, named after Gerhard Damköhler, a little-known German scientist who conducted much of his work during the rise of Nazi Germany.* The chemistry of a flowing gas or liquid involves many processes and phenomena, which means that determining the chemical state of a mixture is a complex problem, the only "reference" state being equilibrium when all the reactions balance each other and end up in a stable composition. Damköhler attempted to work out a relationship for the rate of chemical reactions or diffusion in the presence of a flow. In this non-equilibrium scenario, Damköhler came up with a single number—the Damköhler number—that can be used to compare the time for the "chemistry to happen" with the flow rate in the same region.[13]

The Damköhler number is useful when it comes to the placenta because the organ is diffusing solutes, such as oxygen, glucose, and urea, in the presence of both a fetal and maternal blood flow. Here, the Damköhler number is defined as the ratio between the amount of diffusion against the rate of blood flow. Chernyavsky found that, despite the various complex arrangements of fetal capillaries in the terminal villus, the movement of different gases in

---

* Damköhler, the son of a physician, was born in 1908 in west central Germany. In 1926, he studied chemistry at the University of Munich under the physicist Arnold Sommerfeld. After graduating in 1931 with a doctorate in chemistry, Damköhler moved to Göttingen University to work with Arnold Eucken, who was director of the university's Institute of Physical Chemistry. In 1937, Damköhler was working in Braunschweig at the Motors Research Institute on combustion and how sound propagated in flames when his research caught the attention of the upper echelons of the Nazi Party, who wanted to use the knowledge for the development of the Luftwaffe's jet engine program. Damköhler, however, wanted his research to be free from political influence, and it is speculated that the pressure contributed to him taking his own life in 1944 at thirty-six. Despite still being so young when he died, Damköhler had already made his name in chemistry via several breakthroughs. Damköhler's body of work, originally published in German, remains of high value. After the war, for example, NASA translated his research to help them develop rocket technology.

and out of the fetal capillaries could be described by the Damköhler number, which he calls the "unifying principle" in the placenta.

For a number much larger than one, diffusion dominates the process and occurs much faster than the blood flow rate. In this case, the only way to increase the uptake of solutes—say, in the fetal vessels—is to increase the rate of blood flow. This is said to be "flow limited." For a Damköhler number much smaller than one, the flow rate is greater than the diffusion rate, so the diffusion is slow. Here, the rate of blood flow is so fast that it dominates, and the only way to get through this "bottleneck" is by increasing the diffusion rate. This is called "diffusion limited" exchange.

When Chernyavsky and colleagues fixed a slight pressure difference between the inlets and outlets of the fetal blood vessels to around 0.3 millimeters of mercury (or 0.3 mmHg),* they discovered a huge range of Damköhler numbers—from 0.01 to 100—for different solutes.[14] They found, for example, that carbon monoxide and glucose in the placenta are diffusion limited while carbon dioxide and urea are more flow limited.† Carbon monoxide is thought to be efficiently exchanged by the placenta, which is why maternal smoking and air pollution can be dangerous for the baby. Interestingly, oxygen is close to being both flow and diffusion limited, suggesting a design that is perhaps optimized for oxygen—which makes sense given it is so critical to life. Why there is such a wide range of Damköhler numbers is unknown and is the subject of ongoing research. One possible explanation is that the placenta has many different roles, which include both nourishing and protecting the baby from harm.

The modeling revealed that different parts of the vessels are optimized for various solutes as either flow or diffusion limited. This is another example of

---

* Millimeters of mercury (mmHg) is a standard unit for measuring pressure.

† Others have explored the exchange in the placenta through the lens of a ratio of key characteristics, notably US scientist J. Job Faber and colleagues some sixty years ago; see: Faber, J. "Review of Flow Limited Transfer in the Placenta." *International Journal of Obstetric Anesthesia* 4, no.4 (1995): 230–237.

the robust nature of the placenta that, as Chernyavsky says, comes at a price of being truly optimal compared to other organs in the body. Chernyavsky and colleagues now aim to create maps of oxygen uptake into the placenta to compare healthy placentas with unhealthy ones, with the ultimate plan of incorporating the mathematical models and noninvasive imaging data via ultrasound or MRI to generate simplified models that could be used for clinical diagnosis—for example, risk assessments for conditions such as fetal-growth restriction and preeclampsia.

There is still a lot we don't know about the placenta, and work on the organ has certainly lagged well behind that of other organs, such as the heart and lungs, for a variety of reasons. There are the ethical concerns of carrying out invasive experiments on pregnant women as well as the short time the organ survives outside of the womb. Animal models, which are extensively used elsewhere to study pregnancy (as we learned about in chapter three) and extrapolate to human models, are of little use when it comes to the placenta. This is because, in the animal kingdom, placentas come in different shapes and sizes and even vary in how far into the uterus they attach to the mother.[15] Even simple aspects such as knowing oxygenation levels in the placenta is difficult, with values varying considerably even with similar techniques. MRI and other noninvasive methods are sensitive to deoxygenated blood, but they cannot give absolute values of the oxygen levels in the placenta. That means the only way to study the placenta in detail is when it is delivered after birth, even though it's only possible to do so for a few hours outside of the human body.

Despite these challenges, imaging techniques such as MRI are throwing up some new surprises. In 2020, UK researchers led by physicist Penny Gowland from the Sir Peter Mansfield Imaging Centre at the University of Nottingham scanned forty-four pregnant women and found how incredibly effective the placenta is at distributing oxygen, even to the extremities of the organ. But they also observed tantalizing evidence of "orchestrated" contractions in the placenta itself, which have never been seen before. This behavior was observed in nineteen women, where the placenta and underlying uterine wall contracted independently of the uterus as a whole. This differed from Braxton-Hicks contractions that contract the entire uterus. These contractions, which were about twenty minutes apart, involved a reduction in the area and the thickening of the underlying uterine wall that was covered by the placenta. The researchers also found up to a 40 percent reduction in the placenta's volume. This entire effect led to the uterus growing in volume—a little bit like squeezing a balloon at one end, which makes the other end expand.

The team suspects that this "utero-placental" pump causes the periodic ejection of maternal blood from the placenta. But, at the same time, it does not affect fetal blood circulation, and it also seems not to make a huge difference in oxygenated blood before or after the contraction.[16] Indeed, blood is not just being pumped into the placenta; it is also being drained away via decidual veins that line the uterus. This draining is often overlooked in studies of the placenta, but Gowland and colleagues think that the pump works to avoid the formation of "unstirred" areas in the placenta that would otherwise limit the efficiency of gas exchange. "There must be a physical purpose for the pump," Gowland said. "But we need to do more work to find out what." A placental pump would be a new phenomenon of the placenta, but what this research and the work by Clark shows is that the placenta and uterus cannot be viewed as individual organs; rather, they act as a coupled system that interacts to produce its own characteristic and, perhaps, crucial behavior.

We know the placenta is robust, having to endure being kicked by the fetus, as we discovered in chapter five, as well as having to ride out the huge contractions that drive the baby out of the uterus, as described in chapter six. But like any other organ, functional issues can result in complications. Sadly, around half of stillbirths in the United States each year are due to conditions affecting the placenta.[17] Women are advised not to take certain drugs during pregnancy, given the potential for them to "cross" the placenta into the baby's bloodstream, resulting in harmful effects. Cigarette smoke as well can reduce the organ's efficiency, which, in turn, impacts the delivery of nutrients and oxygen to the baby. Worse still, the placenta can separate prematurely from the uterine wall, in effect disconnecting from the maternal blood supply that is nourishing the fetus. This is called placental abruption, the cause of which is not entirely known but which occurs in around one in a hundred pregnancies.[18]

Such problems have inspired some researchers to develop "artificial wombs," devices that could do the same job as the placenta and uterus but outside of the body. In 2017, researchers at the Children's Hospital of Philadelphia Research Institute removed fetal lambs via C-section from their mother and kept them alive for a record four weeks outside the womb.[19] The feat was performed by putting the lambs in their own "biobags"—what looked like a vacuum-pack bag for clothes storage—containing a liquid-like amniotic fluid that the lambs could ingest. Attached to the lambs' umbilical cords was a machine that provided a continuous fresh supply of oxygen and nutrients while removing carbon dioxide. The fetal lambs' hearts did all the pumping work.*

The biobag supplied the lambs with an artificial environment that not only kept them alive but also allowed them to flourish. The lambs appeared comfortable as they lay on their side and did everything they would have done in the womb, from swallowing and wriggling their legs to opening their eyes (which must have been rather spooky). As the weeks passed by, the lambs even grew a woolly coat and almost outgrew their bags. Once the four weeks were up, their lungs were deemed mature enough to breathe air unaided and so they were "born"—a process that involved carefully opening the bag and disconnecting them from the various machines.

The first batch of test subjects were terminated to allow the scientists to study their brains and organs in detail to examine how well they had developed. With no issues discovered, the researchers replicated the experiment a second time, but this time the lambs were allowed to live. They were even bottle-fed by the team. The age of the fetal lambs, at 110 days' gestation,† was

* For a rather slick video from the team involved, see: www.youtu.be/z7OsEGT9d3o.

† Full gestation for a lamb is about 147 days.

chosen as it corresponds roughly to a human fetus at twenty-three weeks, which is often said to be when a human baby is "viable." Yet, a baby born at this point, even with advanced medical equipment, only has a 19 percent survival rate.[20] If the infant does survive, the potential complications are considerable, with the infant having about a 75 percent chance of developing a disability or some form of illness, such as chronic lung disease, by age three.[21]

The team in Philadelphia, as well as others who are attempting similar developments, say that one day such biobags could be used for humans, but insist they do not want to lower the limit of viability. Instead, they want to create a device that can be used for extremely premature babies—those born before twenty-eight weeks—to allow them to safely mature for three or four weeks in an environment like the uterus. If this were managed safely, as has now been shown for lambs, it could reduce the chance of serious illness developing. Babies born at twenty-six weeks, for example, have a 50 percent chance of having no impairments by age three.

Despite the success of the fetal lamb experiments, there is a long way to go before we ever see Matrix-style "fetal fields" of human babies in biobags across neonatal intensive care units. Testing biobags on humans represents not only an ethical minefield but also concerns about the long-term health implications of doing so. There are also many technical hurdles to overcome, including how to safely connect to the umbilical vessels of an extremely premature infant. Another issue is the need to have a much better understanding of the physiology of the placenta and how it manages to control a wide variety of nutrients and oxygen to and from the fetus. Those breakthroughs will come from a better physical understanding of how the placenta works so effectively.

For most of the twentieth century, the placenta was viewed as somewhat of a curious black—or rather a red and purple—box, but this is now beginning to change thanks to advances in imaging, modeling, and new experimental techniques that are finally lifting the lid on the dynamics of this ethereal and crucial organ. Additional developments in placental imaging will also help us to better understand the transfer of infections from mother to fetus as well as the exposure of the fetus to environmental pollutants. In 2019, researchers found evidence for black carbon particles in the fetal side of the placenta.[22] A year later, another team found the first evidence of microplastics in the organ.[23] "A decade ago, the placenta was [a] niche, curious organ, but not anymore," Chernyavsky said. "The importance of the placenta cannot be overstated. Its effect is felt not just for nine months but lifelong and ultimately even multigenerational."

Perhaps the Irish playwright George Bernard Shaw put it best in his now-famous quote: "Except during the nine months before he draws his first breath, no man manages his affairs as well as a tree does."

FIRST BREATH

At 12:52 PM on August 7, 1963, US president John F. Kennedy and First Lady Jacqueline Kennedy welcomed their third child into the world.* Born following emergency caesarean section at Otis Air Force Base Hospital in Buzzards Bay, Massachusetts, Patrick Bouvier Kennedy was delivered at just thirty-four weeks' gestation, coming in at 2.11 kilograms (4.65 pounds). As with any babies born preterm at the time, doctors knew he could face difficulty adjusting to life outside the womb. Indeed, just minutes after birth, something was not quite right.

The infant soon developed breathing difficulties, and pediatric specialists were helicoptered in to assist. The pediatricians suspected that the Kennedys' son was suffering from hyaline membrane disease, a condition in which a membrane of dead cells and proteins line the lungs, preventing the lungs from fully functioning.[1] Babies with the condition often had hardship shortly after birth, manifested by fast breathing and a fast heart rate. They advised that Patrick be transferred some seventy miles away to Boston Children's Hospital,

* This was Jacqueline's fifth pregnancy, the first resulting in a miscarriage, the second a stillbirth, and the following two being Caroline Kennedy and John Kennedy Jr., born 1957 and 1960, respectively.

which had better medical resources, and so Patrick was rushed by ambulance to the hospital, arriving about five hours after birth.

What afflicted Patrick was a set of circumstances that was all too familiar in the 1950s. Pediatricians were dumbstruck as to why some premature babies seemed initially fine once born but died two or three days later. Particularly puzzling was that autopsies found no residual air in the lungs—they had totally collapsed. Despite pioneering medical techniques at the time, little could be done to aid premature infants who struggled to breathe—a lottery of life and death, it seemed. And it was not just a few babies. In the US alone, over ten thousand newborns died each year in their futile struggle to breathe while another fifteen thousand had similar issues but somehow came through.[2]

The basic reason for the deaths was due to a lack of oxygen, which is essential to life as we would suffocate without it. As we just learned in the previous chapter, the placenta supplies all the oxygen that the baby needs in utero from the maternal blood supply. In a similar manner, the lungs work by bringing in oxygen from the atmosphere during inhalation and transferring it into the bloodstream, while removing carbon dioxide from the body during exhalation.

This oxygen exchange takes place across membranes of balloon-like structures called alveoli. There are around four hundred million in an adult lung and around twenty-four million in a newborn's lungs. The alveoli are covered in blood vessels that enable oxygen to diffuse across the membranes into the bloodstream while expelling carbon dioxide from the body. The alveoli, which inflate and deflate with each inhalation and exhalation, are arranged in bunches—somewhat like the bunch of balloons that float the house in the movie *Up*.

Fetuses begin practicing breathing at an early gestation period, and in doing so, they ingest amniotic fluid into the partially collapsed lungs. But it becomes a different ball game when they need to breathe oxygen from the air. As soon as the birth partner, or whoever it may be, cuts the umbilical cord, the baby is on its own. There are a few processes just before birth that can help prepare the infant for this adjustment, such as uterine contractions that constrict the umbilical blood vessels. This reduces oxygenated blood to the fetus and so elevates levels of carbon dioxide in the blood, which stimulates the respiratory center in the brain to trigger the newborn to breathe.* Still, birth represents a real crisis for a baby, and taking the first breath to inflate the lungs to near full capacity is far from straightforward, even for full-term babies.

The issue preterm babies face is that the lungs are one of the last organs to fully develop and are only complete around thirty-seven weeks—hence the term "full term." In the 1950s, doctors thought that a layer in the lung, known as the hyaline membrane, was behind the difficulties that preterm babies had in breathing. It was thought that the layer could originate from the amniotic fluid or even milk that was ingested by the infant.

But the discovery in autopsies of affected babies' lungs of a protein usually found in blood put the focus back on the baby itself. The disease became known as infant respiratory distress syndrome, or IRDS, and even today, over 5 percent of babies need assistance in breathing during the first few minutes of life.[3] A failure to breathe after birth still kills a staggering eight hundred thousand babies annually, the majority of which are in developing countries that do not have access to advanced medical equipment.[4]

Such advances, sadly, came too late for Patrick and thousands of other babies before and after. Despite some of the best medical teams in the country and following a "desperate medical effort" to keep Patrick alive, he died two days after birth.[5] "The battle for the Kennedy baby was lost only because medical science has not yet advanced far enough to accomplish as quickly as necessary what the body could do by itself in its own time," the *New York Times* noted at the time. The Kennedy death did, however, shine a light on the condition and spurred further research that resulted a few decades later with scientists having a much better understanding of not only what was

---

* Work in 2020 showed that a specific gene is turned on in the brain, triggering the infant to breathe; see: Shi, Y., Stornetta, D.S., Reklow, R.J., et al. "A Brainstem Peptide System Activated at Birth Protects Postnatal Breathing." *Nature* (2020): doi.org/10.1038/s41586-020-2991-4.

going on in the lungs during IRDS but also potential solutions. The explanation for why it happens in the first place had already been some two hundred years in the making.

The English polymath Thomas Young is regarded as the "last man who knew everything."[6] He began his teenage years having already learned over ten languages, including Greek, Latin, and Arabic, and would famously put this love of languages to use later in life to decipher the Egyptian hieroglyphs via the Rosetta Stone in the 1810s. Born in 1773 in South West England, Young was the eldest of ten children, and after studying medicine at the University of Göttingen in Germany in 1801, he was appointed professor of natural philosophy at the Royal Institution. A year later, he became foreign secretary at the Royal Society. In the 1800s, Young made several breakthroughs in physics that remain the mainstay of textbooks today. His most famous being the investigation of the wave nature of light as well as the relationship between the pressure on a body and its associated change in length because of such a stress, which is known as Young's modulus.

Young also focused his attention on the properties of liquids and the weird trampoline-like film at the top of a liquid surface where it meets the air. You can see this effect yourself if you fill a glass with water and look very carefully at the top of the liquid, where a very thin layer seems to appear. This is due to an effect called "surface tension." Deep in a liquid, water molecules are surrounded by other water molecules on all sides, which results in the interactions between the molecules balancing out—giving no net tension. In other words, the molecule pulling above is balanced out by the molecule pulling below, and so on. However, molecules at the very top of the surface do not have neighbors above them—only at the sides or below. This results in the molecules bonding strongly with their neighboring molecules along the surface, creating a "surface film," much like a stretchy elastic sheet. Some animals make special use of this, such as the pond skaters, or water striders, which can literally walk on water, thanks to surface tension.

Young linked surface tension to another fascinating phenomenon of fluids known as capillary action,[7] which happens when a liquid has the gravity-defying ability to flow in narrow space. Plants take advantage of this phenomenon to obtain water from the ground. At home, you can see how capillary action works by rolling up a paper towel into a cylinder and then dipping a part of it in water. The liquid will magically, and very slowly, rise up the paper towel. This property of fluids was of great interest during Young's time, and, working independently from Young, the French polymath Pierre-Simon Laplace* put these observations into a mathematical formulation.[8,†] The result is what is now known as the Young–Laplace equation, which relates the capillary pressure difference between two interfaces, such as water and air, to the surface tension.

A bubble—a liquid–gas interface, that is—is perhaps the best example of the Young–Laplace equation in action. In this case, the pressure difference between inside and outside the bubble is dependent on the size of the bubble and the surface tension. For a given radius, the relationship between surface tension and pressure is proportional, so a larger surface tension results in a bigger pressure difference, and likely bubble collapse. For a fixed surface tension, the radius and pressure are inversely proportional, so if the size of the droplet increases, then the pressure difference drops. If the size of the droplet decreases, then the pressure difference increases.

The Young–Laplace equation also applies nicely to the bubble-like alveoli in the lungs where a thin film of water in the moist tissue of the alveoli meets air during breathing. Because the water molecules on the alveoli tissue exert a stronger force on each other than they do on the air molecules in the alveoli, this creates a high surface tension. In this case, according to the

---

* Born twenty-three years before Young, Laplace also made countless contributions to mathematical physics, astronomy, and statistics and even dabbled in politics, becoming minister of the interior under Napoleon at the turn of the nineteenth century.

† German mathematician Carl Friedrich Gauss is credited with combining the work of Young and Laplace to create a more complete mathematical description of surface tension.

Young–Laplace equation, the pressure difference is so high that it results in alveoli collapse, never having the capacity to expand again.

But we know that the lungs do not collapse, at least in healthy babies and adults, so there must be something working against this from happening. The first scientist to begin unraveling this mystery was the Swiss physiologist Kurt von Neergaard. In 1929, he showed that, rather bizarrely, it took more pressure to inflate the lungs with air than it did with water. Having a background in physics, von Neergaard used the Young–Laplace equation to demonstrate that surface tension at the boundary between lung tissue and air was the reason for this difference.[9] He even measured the surface tension of healthy lung extracts from pigs, finding it to be lower than other healthy tissues in the body.

Fast-forward twenty years, and the 1950s marked a boon in lung research, thanks mostly to the ravages of two world wars that laid bare the deadly impact of toxic gases. The British physicist and physiologist Richard Pattle had gained a reputation as a bubble expert, particularly for his research into "antifoams" that could prevent bubbles from forming in tissue cultures that were exposed to air. He began working at the top-secret Chemical Defence Experimental Establishment on Porton Down in Southern England, where he was tasked with helping physiologists tackle the formation of foams in the lungs when they were exposed to phosgene.

This poisonous gas had been used during World War I as a chemical weapon and was responsible for the loss of tens of thousands of lives. Inhaling the gas resulted in a pulmonary edema—excess fluid buildup in the lungs—that made it impossible to breathe. Pattle was confident that his antifoams would quickly do the trick. But, much to his surprise, he failed to stop the bubbles from forming in the edema fluid. He concluded that the alveoli must be covered in a layer that somehow reduces the surface tension, which made it easy for the bubbles to form and stabilize.[10] He even speculated in his writing that this lining could be behind why some newborn babies have difficulty breathing.[11]

At the same time as Pattle was conducting his research, John Clements, a physiologist working at the US Army Chemical Center in Edgewood,

Maryland, proved the link between this layer in the lung and its function. Clements, self-taught in physics and mathematics, conducted a series of quantitative measurements of the surface tension of lung extracts from rats, cats, and dogs—minced lungs extracts, to be precise. His homemade apparatus was rather crude, but it could measure the surface tension as he compressed and expanded the surface of the minced lung. He found that the surface tension rose when the lung expanded—in effect, stopping it from over-expanding—while it decreased when the minced lung contracted. This not only stopped the lung from collapsing but also allowed it to easily expand again.[12] Clements's findings confirmed the presence of a substance on the surface of the alveoli that reduced the surface tension by a factor of around ten. With a mechanism discovered, it begged the questions: What was this mysterious layer, and what was the link between this layer and preterm babies struggling to breathe? Let's take a look at washing clothes to find out.

A domestic appliance used almost daily by new parents—apart from the coffee machine—is the washing machine. For such a small body, babies produce an incredible amount of drool, vomit, urine, and poo, all of which will somehow be instantly attracted to your own clothes. Indeed, a rite of passage of any new parent is to suffer, at least once, from a cascade of puke down the back when burping your newborn over the shoulder, regardless of what size of muslin you have placed in the baby's way. Of course, water alone is not enough to get your clothes clean; you also need detergent, made up of clever molecules called surfactants, the most widely used commercial one being alkylbenzene sulfonate.

A surfactant has two important jobs that are carried out by different parts of the molecule. The first is that it connects to water molecules preventing them from bonding strongly with each other. This reduces the surface tension of the water, allowing it to spread over a greater area and seep deeper into the fibers of the clothes. The second is that the surfactant binds with dirt and grease. The wash cycle of the washing machine combines the water with the surfactant, which clings to the dirt, while the tumbling process breaks down the dirt and grease into smaller pieces. The rinse stage wipes away the dirt and detergent leaving you with clean clothes. Et voilà.

There is a simple experiment you can do at home to demonstrate the power of surfactants. If you cut out a piece of paper in the shape of a boat and put it in a bath, it should float on top and not really go anywhere. However, if you drop a tiny amount of dishwashing liquid at one end of the boat (before it gets soggy and sinks), it will magically propel away—and at a decent speed. This is because the detergent is reducing the surface tension of the water in a way that breaks this virtual film and provides a reactive force that pushes the boat. This same principle is why soap bubbles, a favorite attraction for any child, can last for so long and do not burst. The surfactant reduces the surface tension in the bubble, and this reduces the pressure difference between the outside and inside of the bubble to keep it in equilibrium—at least, that is, until it gets thin enough to break. Pattle was seeing a similar phenomenon in the 1950s when investigating edema fluids: the surfactant in the lung was stabilizing the foams so they would not collapse.

At the same time as Pattle was conducting his research in the mid-1950s, pediatrician Mary Ellen Avery at the Harvard T.H. Chan School of Public Health (Harvard Chan School)* saw for herself how many babies struggled to breathe following birth. Avery, who worked at the newborn service at a local hospital in Boston, was convinced that surfactants were the missing piece of the jigsaw, and building on the work of Clements, she constructed her own apparatus to measure the surface tension of the lungs of babies who had died of IRDS. For babies without IRDS who had died of other complications, she observed the same effect that Clements had observed: a lowering of the surface tension as the lung collapsed to help it reinflate. For IRDS babies, however, she found that when the lung expanded, the surface tension rose to much greater levels than non-IRDS babies. Worse still, when it was compressed, it remained so high that it would have made it significantly harder to

---

* In 2014, the Harvard School of Public Health was renamed the Harvard T.H. Chan School of Public Health following a donation from the Morningside Foundation.

take another breath.[13] Avery and her colleague Jere Mead, also at the Harvard Chan School, showed that there must be a surfactant in the lung that lowers the surface tension to almost zero.

Initially, their findings were met with skepticism, with some unconvinced that the lungs are chemically active enough to produce such a substance. Yet, decades later, imaging of the lung cells by powerful microscopes showed that they do indeed contain cells that produce pulmonary surfactant. The alveolar epithelial layer was shown to be made up of two main cells. The first, dubbed Type I, covers around 90 percent of the air-sac surface and is responsible for gas exchange, with oxygen and carbon dioxide passing through these cells to and from the bloodstream. Type II cells, which cover the remaining 10 percent, are responsible for producing the surfactant.

About 92 percent of Type II cells' composition is made up of lipids, with the remaining 8 percent composed of proteins.[14] Despite making up a much smaller percentage of the surfactant, these proteins in the Type II cells are incredibly important for lung function. Two of them are hydrophobic, meaning that they do not like water. This helps drive the surfactant to the air-liquid surface where it can be effective. Another two were initially harder to decipher but were found to be related to other proteins that are critical for the immune system. It is thought, therefore, that the surfactant is also important for stimulating immune responses in the lungs and may be responsible for protecting the lungs against infection. This surfactant layer in the lung is so effective that it is possible to inflate the alveoli with only 1 mmHg of excess pressure over the surroundings.* Yet, analysis of infants at different gestation periods shows that surfactant production only begins around week twenty-four. At about thirty-four weeks' gestation, babies may have enough surfactant to breathe on their own before production reaches its full level by thirty-seven weeks.[15]

All this knowledge of what was happening in the lung offered a chance to finally put IRDS behind us. In the 1950s, pediatricians put babies with IRDS on ventilators that could supply a pressure during inhalation to help inflate the alveoli sacs. This was a high risk, however, given that it can cause long-term

* In comparison, atmospheric pressure at sea level is 760 mmHg.

damage to the fragile lungs. The work of Avery and Mead showed that only providing pressure when the baby is inhaling is not enough, and pediatricians began using another device, called the continuous positive airway pressure (CPAP), which works like the missing surfactant by maintaining a pressure even when the baby breathes out. CPAP is somewhat like sticking your head out the window in a moving car and opening your mouth. In a clinical setting, it initially involved fitting a mask over the baby's nose and mouth, and it was incredibly effective, reducing the mortality rate from 80 to 20 percent.[16] It was so successful that it was introduced without clinical trials where it is now given via a nasal cannula.

Once a surfactant was determined to be the main factor behind IRDS, scientists began to think about ways to tackle the issue directly. Many had labored for years to develop replacement surfactants, and while they achieved some success in rabbits, humans proved more difficult. A breakthrough came via the Japanese physician Tetsurō Fujiwara, who in the late 1970s visited a local slaughterhouse to obtain a fresh supply of cow lungs from which he extracted the surfactant. After washing and mixing it in solution, he administered it into babies' lungs through a tube put directly into the windpipe. This involved placing a tube down the baby's throat and pumping the surfactant, looking similar to skimmed milk, into the trachea, or windpipe. In 1980, Fujiwara gave it to ten premature babies who were aged between twenty-eight and thirty-three weeks and saw a huge improvement in their chest X-rays, with eight of the ten infants surviving.[17]

Some pediatricians, however, were uncomfortable putting cow-lung extract into the lungs of premature babies, and so scientists began to create synthetic surfactants made in the lab. Following a decade of clinical trials in the 1980s with both animal-derived and synthetic versions of liquid replacement surfactants, they were approved for clinical use.[18] All these developments led to a rapid decrease in the number of deaths from IRDS, from around 12,000 in the United States alone each year to fewer than one thousand in the early 2000s.*

---

* Experiments also showed that steroids could accelerate lung development. Lambs that were born around a month early suffered a quick death from respiratory distress, but

Despite that progress, once a preterm baby is born today, it can still take several hours to assess the severity of breathing. Babies struggling to breathe are usually first treated with CPAP. But if their oxygen levels remain too low, they may be given surfactant replacement therapy. CPAP also is not without issues. As noted before, it can damage delicate lung tissue in preterm infants. And, currently, surfactant replacement is no panacea. The procedure is not patient specific or tailored to the geometry of individual lungs, and as several rounds of treatment may be needed, it increases the risk that excess fluid builds up in the lungs or that the lungs become blocked altogether. It is also impossible to monitor how effective the treatment is in real time. Instead, analysis takes place through lung-function tests and blood-oxygenation levels after treatment, or, rather crudely, just seeing whether the infant gets better in time. Worse still, about 37 percent of babies do not respond to treatment at all.[19]

All of which means there is still room for much improvement.

Mehdi Raessi at the University of Massachusetts Dartmouth has been interested in the problem of surfactant delivery in infants for over a decade. His biggest challenge, however, was not the difficulty of studying the problem but rather obtaining funding to do so. It took him years to persuade the powers that be at the National Science Foundation—one of the main US funders of basic research—to support his group's proposals. His group was finally awarded a grant in 2019. Given that the flow of surfactants is tricky to determine experimentally in situ, his team took to computational models to simulate the surfactant flow through the lungs. It sounds simple, but it is far from trivial given the fluid mechanics of the surfactant liquid plug and the sheer number of bronchial tube divisions in the lung that need to be navigated.

---

sheep treated with steroids could deliver preterm fetuses that were able to breathe on their own. Obstetricians found the same effect in humans. Beginning in the 1990s, steroid treatment for IRDS became widespread. The treatment consists of administering a steroid injection in pregnant women just before the baby is born to help accelerate the development of the lungs.

The lungs are attached to the trachea, or windpipe, an elastic tube that splits like an upside-down Y into two primary bronchi, each one reaching into a separate lung. Each bronchial tube splits further into more and more generations of Y-shaped tubes, like a tree. The number of divisions reaches about eight in the newborn lung (compared to some fifteen in the adult lung). The final tubes that lead to bunches of alveoli are the bronchioles, which are about 1–2 millimeters (about the size of a period) in diameter.

As the plug (i.e., surfactant layer) moves through the lung, it coats the airway walls and can also split into two at the numerous generations of Y-junctions in the lung that get progressively smaller. Eventually, the plug will hopefully reach the bronchioles and drain into the alveoli sacs, which are each around 150 micrometers in diameter but can be half that for premature babies. The problem with surfactant treatment is that it must reach the extremities of the lungs—there is little point in it solely lining the bronchial tubes on the way there—and also needs to reach as many different alveoli as possible. If it all drains into the same place, then the treatment will be of little use.

Theoretical models in 2015 showed that about half of the plug ends up lining the lungs as it moves, while the other half is delivered into the alveoli.[20] Researchers also found that increasing the flow rate results in more uniform splits, but the coating rate also increases, thus lowering the amount of the plug transported into the alveoli. Yet, simply ramping up the flow rate is difficult in preterm babies, as it could damage the infant's lungs. The issue with these previous models is that they treated the lung as purely symmetrical. But lungs are not perfectly uniform, instead having asymmetries both in the size and direction of the branches, which could become a big factor in the infant lung. Indeed, experimental work in 2019 on the rat lung—which is similar to the human lung in terms of the ratio between the diameter and length of the bronchioles and is also asymmetric—found that the surfactant was poorly distributed.[21] Another problem in surfactant distribution is that some Y-tube splits turn "upward," (i.e., against gravity). This makes it more favorable for the fluid to drain "downward," preventing the fluid from distributing across

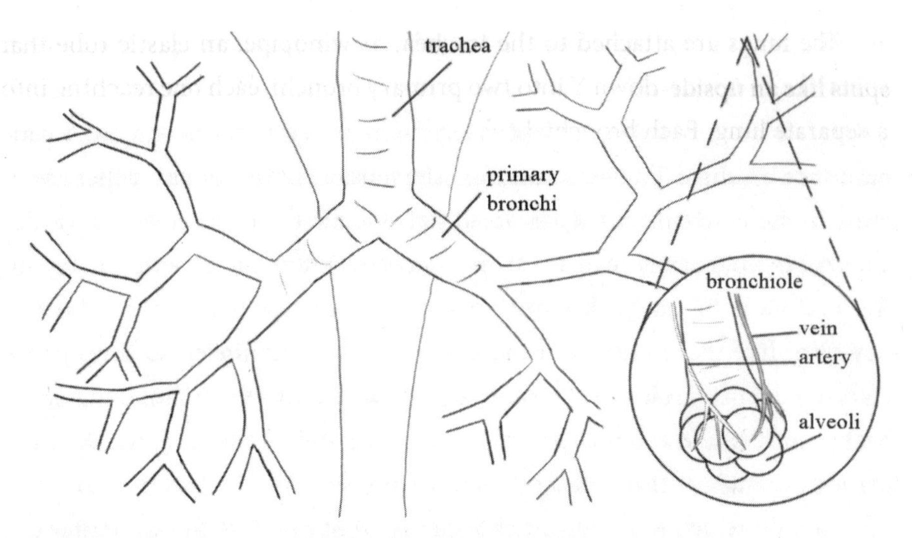

Fig 1 | *Schematic of the lung.*

the whole lung. This effect is even more pronounced in adult lungs, which are much larger.

Raessi's team, including postgraduate student Cory Hoi, wanted to understand what parameters are important in delivering the surfactant to the alveoli—be it the viscosity of the surfactant, how fast it is delivered in the lung, or even the timings of multiple surfactant doses. They first studied the mechanics of the surfactant flow along both two and three generations of Y-shaped capillaries and how it splits at the various junctions. We know that capillary action can help fluids travel against gravity, just as it does in plants and vertical paper towels. The researchers, therefore, thought there may be a way to simulate this by manipulating how often surfactant is pumped into the lung. They found, at least in their 3D models, that the timing of the dose can help. When one surfactant dose goes down one side of a Y-split due to gravity, if it remains in that section of capillary when another dose is injected, the new plug will generally enter the other part of the Y split—even going against the effect of gravity. This means that if an initial dose of about 10 millimeters in length were instead given in two 5-millimeter doses, the method could help to produce a more distributed surfactant dose rather than draining toward one way.

This is fine in a two-generations lung airway tree, but of course we do not have such a simple lung structure, so Raessi has teamed up with experimentalist Hossein Tavana at the University of Akron. The pair are now creating a more physiologically realistic 3D replica of a six-to-eight generation airway tree.[22] For the moment, the work is ongoing, but that is not to say further down the line the results from these models could be applied so simply in a clinical setting. Surfactant is usually administered with the infant lying on its back, which would promote the surfactant to travel in one way. But the simulations may show that rolling the baby from side to side during treatment, for example, which is not included in current best practice, could counter the negative effects of gravity. Other possibilities include splitting the plug up into doses given at more specific set times or perhaps playing with the viscosity of the plug itself. "We are very hopeful that this more realistic model will shed some light on the issue of surfactant distribution," Raessi told me.

A surfactant that could be safely and effectively distributed could help bring infants off CPAP faster. But perhaps the ultimate aim for infants who are nearer to full term might not involve CPAP or surfactant replacement therapy at all. Just as scientists are investigating biobags to help keep very premature babies in an environment as close as possible to that of the uterus (see previous chapter), scientists in Canada and Germany have been building an artificial lung that could, in a similar way, keep a baby alive until its lungs begin to fully function.[23] Mechanical engineer Ravi Selvaganapathy and his colleagues at McMaster University in Canada teamed up with pediatricians at the same university to build a device from scratch that could oxygenate the blood without the need for CPAP ventilation or surgery, which is required in techniques like extracorporeal membrane oxygenation.*

After a decade of building several generations of devices, they created one that is portable, requires no external power, and does not need an external

* Known as ECMO, this technique involves pumping the blood through an oxygenator that is outside of the body and, of course, is highly invasive, requiring surgery.

supply of oxygen—no easy feat. About the size of a microwave oven, it connects to the umbilical cord, using the heart to pump blood, which is advantageous as the body controls the flow of blood rather than an external pump. The device is based on microfluidics, which involves the behavior and control of small amounts of fluids through channels that are micrometers in diameter—usually sets of small channels etched or molded into a material, such as glass or silicon. For this device, the team used a silicone-based organic polymer called polydimethylsiloxane, or more simply PDMS, in which the small channels can be easily made via a process called soft lithography.*

The deoxygenated blood is pumped by the heart into the device at a pressure of about 30 mmHg, the same as that in a newborn. The blood then passes through small channels—around 100 micrometers in diameter or about the thickness of a human hair—that increases the chance of diffusion of oxygen into the blood and carbon dioxide out. These channels are contained in a series of sixteen microfluidic membranes each about the size of a small dinner plate, which are arranged in a series, like plates lined up in a dishwasher—an arrangement that also increases the surface area for exchange between the blood and the air. This stacking system also has the advantage that it is possible to add or subtract modules depending on the weight of the baby, a bit like having a bigger artificial lung for larger babies. By the time the blood emerges, it is oxygenated and then goes back into the umbilical cord to the baby.

According to Selvaganapathy, the device is the first time that a microfluidic artificial lung has been used in conjunction with the natural pumping of the heart. They tested the operation of their device on a live newborn piglet, similar in weight to a newborn human baby and with a similar blood volume. The piglet was born at term. To stimulate IRDS, it was given carbon dioxide for a while before being put on the machine. The team was able to restore the piglet's blood-oxygen levels from about 75 percent to 100 percent. The team's latest device is the tenth iteration. The first device could oxygenate

* Perhaps the biggest application of microfluidics, and in some sense what kicked off the technology in the 1980s, are inkjet printers, which can precisely position dots of liquid that are smaller than the width of a human hair.

about 8 milliliters (just under a quarter of a fluid ounce) of blood in a minute; the latest model can do about 100 milliliters (about 3.5 fluid ounces) in that same time.

The device is at the proof-of-principle stage. The next step is to demonstrate that the device can be used to sustain oxygen supply in a piglet for three to five days and then expand this to testing twenty piglets. If the results of these experiments are positive, then the device, with improvements, could go on to clinical trials. If this is successful, then the team is confident that the device may be available for clinical use on preterm babies by the end of the 2020s. Yet, there are several technical hurdles that will need to be overcome. Given that the umbilical vein can collapse a few minutes after being cut, an aspect that needs further work is testing the ability to insert a catheter into the umbilical cord to keep it expanded. (The experiments on the piglets were done via the jugular vein.)

The device is designed to handle about 10 milliliters (just over a quarter of a fluid ounce) of blood—or about 10 percent of a newborn's total volume—at any given time so as to not dilute the baby's blood when the device is hooked up. That means that the baby will need to breathe to some extent on its own. The device will likely be able to supply about 30 percent of oxygen, with the remaining 70 percent coming from breathing. But if more support is needed, then additional modular units could be added. The team expects that the method could be used on preterm babies born around thirty-six weeks' gestation, but not any sooner given that the newborn could only be on the device for a week. Although the work is far from done, these new techniques and models promise to make treatment for preterm infants with breathing difficulties safer and more effective than ever.

The first breath of a healthy infant is usually taken within ten seconds after clearing mucus from the mouth and nose, while the remaining fluid in the lungs drains away or is absorbed by the body in anywhere from ten minutes to four hours.[24] Once the lungs fill up with air, instinct should kick in, and the baby lets out an ear-piercing cry, marking the start of healthy respiration. While that is challenging for those born preterm, thanks to the theory of surface tension developed by Young and Laplace some two hundred years ago, as well as the pioneering work of physicians in a magical decade of discovery in the 1950s, the lives of millions of babies around the world have been saved. The discovery of lung surfactant is one of the most powerful examples of a problem identified in patients, understood by scientists, and then targeted with treatment options in hospitals. Thanks to those advances, a baby born today at the same gestation as Patrick Kennedy has a near 100 percent chance of survival.

10

THE SOUND OF CHAOS IN NEWBORN CRIES

A newborn's cry is a distress call like no other, like a sledgehammer to the brain in the middle of the night while still in a sleep-deprived state from the night before. The howls of a very hungry baby will be waking you up every night for some time to come (sorry!).

I have experienced my own baby's midnight shrills, and it was not unlike the hair-standing vocals of the lion I heard in a too-close-for-comfort encounter during a safari on the outskirts of Kruger National Park in South Africa, where my wife and I ventured for our honeymoon. During an early-morning drive, reports came through via the radio that two nomadic male lions were fighting with a male member of the local pride. Our driver drove off at great speed, skidding on the track and throwing his passengers around (no seat belts in the savannah). When we caught up with the lions, however, it was too late—the two nomads had made the kill.

The driver decided to slowly follow one of the nomadic lions as it walked off. We moved slowly behind, tracking its every movement. The lion then turned off the beaten track and, rather surprisingly, we still followed. The grassy terrain began to get bumpier and suddenly, the lion stopped and turned to face us. Looking directly into our eyes, the lion let out one of those roars that resonates through your whole body, making every hair stand on end. It was without doubt the most terrifying moment I have ever witnessed—apart, that is, from the time we took the removable sides off the cot bed for our two-year-old. The lion was making a simple statement to exert its dominance—don't even think about coming closer.

What I took away from the episode was the spine-tingling vocals that a lion can produce. They are able to generate up to 114 decibels at one meter.* This intensity is around twenty times louder than a lawn mower at the same distance and is not far from the so-called "threshold of pain," which is about 120 decibels. Human babies can't quite match the same sound levels as a lion, but they don't come too far from it at up to around 100 decibels. The problem is that, unlike a lion, you are usually holding a newborn fairly close to your ear when the child lets it rip. And while it might damage your ears if too close, the baby will be fine. All mammals, including lions, have an inbuilt volume control called the acoustic reflex. Once the brain starts to kick-start vocalization, it contracts the tensor tympani and stapedius muscles—located in the middle ear—that act to reduce the ear's sensitivity. And it is a good thing that babies have this mechanism given how much they cry.

The saying goes that newborn babies cry for two hours every two hours. That is not true, of course, but they still do a lot of it. A study of 8,700 babies in 2017 found that newborns cry for a total of two hours per day, on average, dropping to just over an hour by three months.[1] All babies are different—some may cry for hours on end for seemingly no reason (sometimes referred to as colic), while others might not reach such highs, or lows as the case may be. But what is consistent among newborns is how full of emotion their ear-piercing shrieks are. The wails are known as "body near." In other words, they express a physical condition, such as hunger. This has prompted researchers to seek ways to analyze the sounds that infants produce to deduce whether they can

* The decibel, or dB, is the unit of measurement for sound level. Sound spreads out in all directions as it travels, so the energy gets spread out over a sphere as it travels away. The farther away you are, the lower the intensity—and sound level—of the roar.

be "translated" into cries of either hunger, hunger, or more hunger—actually, hunger, pain, or frustration.

To anyone within earshot, these emotional newborn cries have been shown to have a powerful effect on the brain—a response that occurs across the animal kingdom. In 2014, biologists in Canada and the United States conducted field studies of deer in which they played the cries of a fawn through loudspeakers. They discovered that the mother deer almost always came running toward the source of the sound.[2] The same behavior also occurred when the researchers played the cries of various other animals—baby bats, sea lion pups, kittens, and a kid goat—but changed the pitch, or the perception of the sound, to that of a cry produced by a fawn.

A study in mice a year later in 2015 found that upon hearing a lost pup's distress cries, a female mouse retrieved her pup by the scruff of the neck, but this had to be learned—a first-time mother initially was not attuned to the cries but got better as time went on. When the scientists then delivered oxytocin to the auditory cortex, the region involved with hearing and interpreting sounds, in pupless females, even they began acting like experienced parents, retrieving the pups when they had never heard the cry before.[3] In 2017, researchers from Germany, France, and the United States took the work a step further and removed a cluster of brain cells—about 17,000 neurons—responsible for producing fast, active breathing in mice pups. The result: when these pups opened their mouths to cry, nothing audible was produced.[4] Incredibly, the mothers totally ignored all the pups and they died, showing the powerful impact that sound has on the early care of offspring.

But what about humans? Do we "suffer" the same effect? It seems so. In 2017, researchers studied the brain patterns that are associated with new mothers' responses to infant cries. They studied 684 women from eleven countries, including Argentina, Belgium, France, Israel, Japan, Kenya, and the United States. The study concluded that, regardless of the country, mothers almost always picked up their crying babies and talked to them.[5] Carrying out MRI scans on some of these women, they found that infant cries activated similar brain regions in both new and experienced mothers. One region that lit up in the brain scans was the supplementary motor area, which is associated with the intention to move and speak, while the inferior frontal regions associated with speech also showed activity.

Although women are undoubtably affected by infant cries, the same cries also affect men, but perhaps not as much. A study in 2013 subjected eighteen men and women to a hungry baby's cry.[6] In female brains, the sound interrupted normal brain activity, jolting the women into action. Yet, the effect was not seen to the same extent in male brains. Which, at least, goes some way to explain how I was never disturbed in the night when my sons were crying with hunger. My wife was the one who either slumped out of bed to give them what they needed or gave me a shove when I had to do it. Well, at least I now have my scientific excuse.

Part of that lurch into caregiving—at least for some of us—is due to how loud babies can cry. After all, nobody wants to listen to those cries for any longer than is absolutely necessary. Babies are not only incredibly good at producing noise, in every sense of the word as we shall soon see, but they can sustain it for long periods. Humans, as well as all mammals, produce these dizzying sounds via the vocal cords, also known as vocal folds, located in the larynx, or the voice box. Thanks to evolution, all mammalian larynxes work in the same way. The diaphragm pushes air from the lungs into the trachea, at the top of which is a 5-centimeter-long (about two inches) tube called the larynx. Roughly in the middle of the larynx are two vocal cords, each appearing like a flap of skin stretched horizontally across the larynx, similar to two drawn window curtains. The pressure of the airflow from the lungs causes the vocal cords to vibrate like "reeds" to produce sound waves.* Nothing in the human body oscillates with a higher frequency than the vocal cords, and the sound produced by them depends on the vocal cords' length and elasticity.

---

* After the air passes through the vocal cords, they close thanks to a principle known as the Bernoulli effect—named after the eighteenth-century Swiss mathematician Daniel Bernoulli.

Given that babies can keep crying for long durations without letup, how can they do so without damaging their vocal cords? If you spent most of your time screaming at the top of your voice, then your vocal cords would not be the same the next day. In this regard, babies have more in common with lions and tigers than you might think. In 2011, Ingo Titze,* director of the National Center for Voice and Speech at the University of Utah, who is one of the world's leading experts on the mammalian voice, and his colleagues dissected tiger and lion larynxes and subjected their vocal cords to mechanical tests to understand how much strain the tissues could take. They found that only a little amount of lung pressure was required to make large vibrations and an intense sound.[7] Another feature Titze discovered is that lions also have a layer of fat within the vocal folds, where some other animals have a ligament. Fat is squishy, offering more leeway for the vocal cords to vibrate. Titze and colleagues suspect that the fat cells inside lion and tiger larynxes are also able to repair quickly so that the cords do not suffer damage.

A newborn's vocal cords are only around 7 millimeters (about a quarter of an inch) long, but they are solely composed of a temporary monolayer that has a homogenous compact collagen structure, a bit like a gel. This gel-like layer inside their vocal cords provides extra cushioning, which is like the layer of fat in a lion's vocal cords. This single-layered structure can be driven very hard by the lungs without suffering injury to allow infants to sustain long—and loud—periods of crying. It is thought that infants can repair damage to the vocal cords around ten times quicker than adults. On top of this, babies also have a thick layer of mucus that helps to absorb energy, with the molecular structure of the mucus aiding in rapid repair. As the infant ages, this single

---

* Titze is also an accomplished tenor and has carried out pioneering research on the operatic voice. He famously sang "Nessun Dorma" from Giacomo Puccini's opera Turandot alongside a computerized version of a human larynx. The computer appeared as the famous tenor Luciano Pavarotti and no recordings of his voice were used, but rather the general anatomy of a tenor's voice was an input to a model that produced the sounds. You can view it here: www.youtube.com/watch?v=CE6zy8aUwtQ.

layer gets thinner and thinner until by the time the child is around three or four, the vocal cords have an adult-like structure.*

Making a noise with the vocal tract, known as phonation, is a complicated process that involves many different aspects of the vocal tract. The acoustic wave created by the vocal cords propagates through the throat, nose, and mouth to produce the sound of your voice with a distinctive pitch. Of course, it is more complicated than that and can be tweaked as it passes through. While the nasal cavity is a fixed size, the tongue and lips can change the size of the oral cavity to change the properties of the acoustic wave. For example, if you say out loud "aaaaaaaaaaaa," "ooooooo," or "iiiiiiiiii," you will notice your mouth moving to produce the required vowel sound. All of these factors—the size and shape of the vocal cords and the size and shape of the throat, nose, and mouth—determine the tone of voice. It's like the vocal cords are the source of the sound and the vocal tract filters it.

The usual way for humans to speak, cry, or sing is for the vocal cords to oscillate in perfect synchronicity, leading to regular oscillations. The lowest and strongest natural vibration of the vocal cords is called the "fundamental" frequency. Men typically produce a fundamental frequency between 85 and 155 hertz while women can go a bit higher due to their vocal folds being a bit shorter (frequency is inversely proportional to vocal-fold length). In addition to the fundamental frequency in a phonation, there are also other frequencies that are at integer multiples, or 1x, 2x, 3x, and so on, of the fundamental frequency. These are known as harmonics, so a 200 hertz fundamental frequency, for example, will have harmonics at 400 hertz and 600 hertz but at lower intensities than the fundamental frequency. Indeed, the vocal-cord system is a perfect manifestation of a complex system having complex inputs

* The vocal cords are around 20 millimeters (about 0.8 inches) long in men and about 14 millimeters (about 0.6 inches) long in women. The vocal cord has an outer layer—the epithelium (or skin)—that is around 0.1 millimeters thick and covered in mucus, while the next layer, the lamina propria, is composed mostly of elastic fibers, fibroblasts and collagen fibers. The third layer—the thyroarytenoid muscle—forms the body of the vocal cord.

but somehow self-organizing to produce, in many cases, a simple output—a fundamental frequency with related harmonics.

Since the mid-twentieth century, this perfect synchronicity—fundamental and related harmonics—was thought to be the primary way to produce sound with the vocal cords. But it turns out that baby humans, and indeed other mammals, do something much more interesting. They turn to rough, or "chaotic," acoustics, especially when turning it up to 11, which could go some way to explain why a newborn cry has such a powerful impact on the brain. So, what do we mean by chaotic phonation and how and why can newborns produce it?

In the early twentieth century, the classical view of the world was that physical phenomena could be characterized into a simplistic dichotomy—determinism and randomness. A deterministic system follows a pattern that is predictable and regular while a random one shows no such pattern, like rolling a dice. In other words, things were either predictable or they weren't. That changed dramatically in the early 1960s when the American mathematician and meteorologist Edward Lorenz showed there is something that manages to combine both these aspects—now known as deterministic chaos. Lorenz was studying weather forecasting and using complex calculations to carry out long-range forecasts. He discovered that his equations were not purely deterministic, but were very sensitive to changes in the initial state of the system. Indeed, even a tiny difference in the values used in the forecasts could have a dramatic long-term effect. Lorenz's work sparked a scientific revolution, and he became known as a founding father of modern chaos theory—the study of apparently random behavior in systems that are, in fact, governed by deterministic laws.*

This sensitivity to the initial conditions—a key aspect of a chaotic system—is known more widely as the butterfly effect, and the best and most simple explanation I have seen for this in action is in the classic film *Jurassic Park*.

---

* For everything about chaos and more on this story, see the classic popular-science book *Chaos: Making a New Science*. Gleick, J. (1987) Viking Books (New York).

One of the film's most enduring, and endearing, characters is Ian Malcolm (played by Jeff Goldblum), a mathematician who specializes in chaos theory. He is invited by insurance lawyer Donald Gennaro to highlight any potential problems with John Hammond's dinosaur theme park, Jurassic Park. Malcolm is pessimistic about the concept of the park, and throughout the film, does not stop short of offering his feedback (and, of course, is proved right as the film progresses). In one scene, Malcolm is in a car on a guided tour of the park together with the paleobotanist Ellie Sattler (played by Laura Dern), who has also been brought to the park to cast her judgment.

Malcolm begins to talk about his work—chaos theory—all the while flirting with Sattler. Malcolm takes Sattler's hand and gently drips a drop of water onto the back of her hand, and they both watch as it trickles away. He then takes a second drop and does the same, finding that it falls off in a different direction. "And there it is," he said. What Malcolm was alluding to is that any slight change in the initial conditions (in this case, a slight change in where the drop lands on the hand) can lead to drastically different outcomes.

Other scientists around the same time as Lorenz were independently discovering the richness of chaos. One of the best-known examples in nature was found in the 1970s by the Australian theoretical physicist Robert May. He was a pioneer of theoretical biology and used his mathematical background to study problems in ecology. While at Princeton University, May investigated a mathematical formulation for how populations could boom and bust. He found that as he increased a certain parameter in the equation, the population rose in a simple way, a bit like a straight line in a graph. So far, so good, but at a certain critical point, the output from the model would not stabilize and would jump between two values (see figure). It was like the number of fish in a pond oscillating between two different populations in alternating years: the formulation did not want to fix on one particular population. This flip-flopping from one regime to the next is known as bifurcation and is a key aspect of a chaotic system.

If this critical parameter was cranked up even more, then the system reached another point with a split to four different population sizes, and so

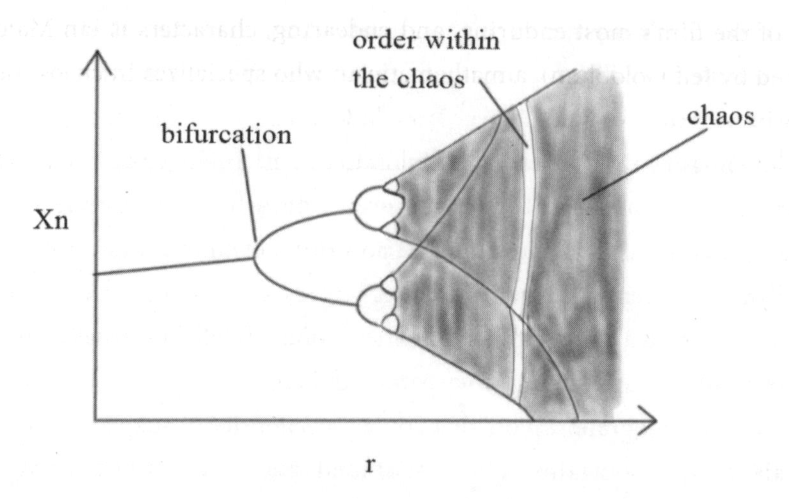

Fig 1 | *Changing a single parameter in a nonlinear system can have a chaotic effect.*

on, until eventually this periodicity between values gave way to chaos in that it jumps around many, many times and never truly settles at all. Interestingly, as the parameter was increased while still in the chaotic regime, more stable regimes lurked where the population came back to being stable between two values—there were windows of order within the chaos.[8] It was an incredible finding that helped to create the field of theoretical ecology.

Scientists had a new toy model to play with, and they found evidence of chaos in most places they looked. It even made physicists reexamine old topics they thought to be mostly solved, such as the simple pendulum. The world is full of pendulums, especially if you visit your local park (see this book's fourth interlude). A simple swing is an oscillator that can sustain periodic behavior, but as every pesky teenager knows, if you push a swing too much, at some point the chain will buckle and there is a "discontinuity" in the motion. Similar issues and exotic behavior can be seen when you have more than one oscillator that are coupled together. Research in the 1970s showed that a supposedly simple model of two coupled oscillators can produce complex patterns of motion, including chaotic behavior. Coupled oscillators are seen in clock pendulums mounted on a wood structure (coupled via the structure itself) and in the rhythm of the heartbeat, where chaos can be seen in irregular heartbeats (slightly worrying to think of such behavior in a vital organ like the heart).

You might be asking what all this has to do with infant cries? Well, another example of a coupled oscillator is the vocal cords, which are subjected to complex "inputs" such as the pressure and flow of air from the lungs. But it wasn't until the late twentieth century that signatures of chaos were noticed in the calls, or cries, of animals. When scientists conducted field studies of animals in their natural environment, they recorded their cries and coos for behavioral analysis via sonographs, plots that show the intensity of sounds produced at certain frequencies against time. One animal that was extensively studied was the rhesus macaque (*Macaca mulatta*), especially a population on the island of Cayo Santiago in Puerto Rico.[9] In the 1960s, scientists made numerous recordings, particularly the quick successive calls that were produced by an adult male. The scientists found a range of behavior, including a "clear call," in which one fundamental frequency, and its associated harmonics, dominated. They also recorded other calls, which were still marked as coos, but were slightly rougher in pitch. As time went on, when recording certain individuals, the researchers found that the coos became even rougher, displaying a whole range of frequencies, somewhat like the noise on an old-fashioned TV set when it is turned to the wrong frequency.

Scientists often measured all three of these scenarios in successive calls—the first being a fairly simple fundamental and harmonic frequency, followed by the presence of other harmonics and then "noise"—in which many frequencies are present in the call. But this noisier call was not just a one-off—it featured in around 30 percent of an average individual's sample. Such calls were quite rare in adult males but much more prominent in younger males and females. At the time, biologists focused on the simpler-sounding calls and ignored these noisier features, perhaps because they did not have the necessary tools to analyze them in detail. It was not until the 1980s and 1990s that these types of calls were studied using deterministic chaos, and scientists discovered that there was a certain structure in the noise. It is now thought that such chaotic or rough calls have several functions in the animal kingdom. One is to get attention while another is to give a sense of vulnerability. A third is to impress mates to give a sense of "fitness."

But it wasn't just rhesus macaques that showed these spectrographic features. In 1986, the German theoretical physicist Hanspeter Herzel came across an intriguing PhD thesis on infant cries by Kathleen Wermke, who earned her doctorate in biology in 1987 from Humboldt University in Berlin. Given that the only way a newborn can communicate is through their voice, Wermke wanted to find out whether it was possible to use the cries as an early diagnostic tool for certain neurological conditions, and to do so, she took several spectrograms of newborn cries. At the time, Herzel had just graduated with a PhD in nonlinear dynamics and chaos at the same university. When he saw the spectrogram plots in Wermke's thesis, he was surprised that they seemed to show complex behavior, similar to the recordings that had been made of macaques decades prior. "When I saw the data, it fascinated me," Herzel said. "I jumped at the chance to work on it."

With a theoretical physics background, Herzel did not know much biology and the physiology of the vocal tract, so he spent the next few years in the university library reading about vocal anatomy and voice production. He then teamed up with Wermke and mathematician Hans Werner Mende from the Academy of Sciences in Berlin to work on infant cries. They took high-resolution spectrograms of seventy infants, both premature and full-term babies, who were between one and five days old. Initially, the spectrograms showed the presence of a fundamental frequency together with its harmonics, but as time went on, the cries became rougher and, as Herzel suspected, chaotic.[10] Using the mathematical techniques of chaos theory, they noticed a "remarkably sharp transition" to subharmonics, when the fundamental frequency and its harmonics are present but also extra frequencies that are half the value of the harmonics. There was then another transition into

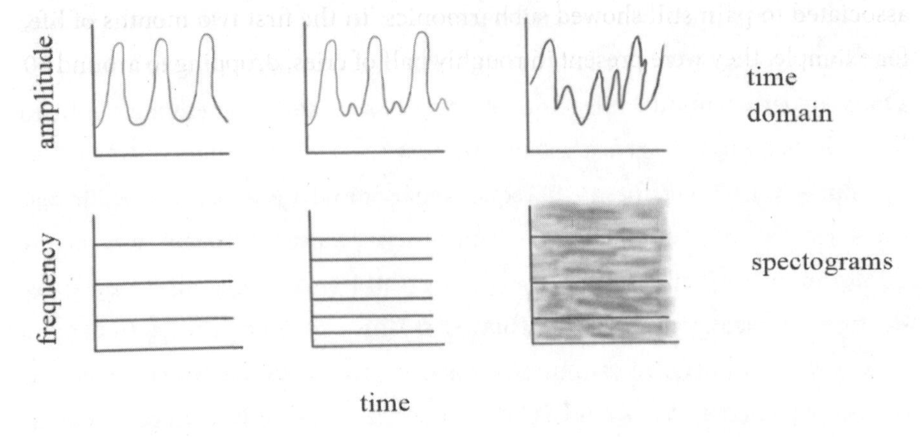

Fig 2 | *Appearance of harmonics (bottom left), subharmonics (bottom middle), and chaos (bottom right) in spectrograms.*

deterministic chaos—in which many frequencies are present—just like with the macaques.

Herzel's result, published in 1990, kicked off the field of nonlinear-human-voice research, resulting in a huge body of work throughout the 1990s. That decade, Titze teamed up with Herzel to develop a model of the larynx and then applied it to people who have vocal-cord paralysis, which also shows chaos.* Physics, it turned out, could be well applied to the output of the larynx, given that it is mainly two muscles that control the pitch of the voice. While the findings were initially unexpected, it is now accepted that infants produce cries that show bifurcation and chaos. And infants use complex phonation regimes more than you might think. Much of the early work on infants concerned cries associated with pain, showing that around half of "pain" cries had subharmonic or chaotic behavior.[11] Later research on cries that were not

* In this case, for example, one vocal cord might go through two oscillations before the other one completes one. Now the whole system takes twice as long to complete the cycle so the period doubles and the frequency halves (frequency being the inverse of period)—producing subharmonics; see: Herzel, H., Berry, D., Titze, I., et al. "Nonlinear Dynamics of the Voice: Signal Analysis and Biomechanical Modeling." *Chaos* 5 (1995): 30–34.

associated to pain still showed subharmonics. In the first two months of life, for example, they were present in roughly half of cries, dropping to around 30 percent at three months.[12]

It is not known exactly where the nonlinear effect comes from in the human voice. It could come from the coupled oscillation of the vocal cords, the stress and strain of the vocal cords, or the pressure of the air as it passes through the vocal cords. "In the body, very little is linear—everything has a nonlinearity to it," Titze said. It begs the question why newborns might produce such noisy and chaotic phonation. In principle, all humans can generate chaotic phonation, but adults tend to "learn" not to produce it mostly because it takes huge lung pressures to drive the vocal cords that hard. For healthy infants, it is not that they lose the ability to produce chaotic phonation, but rather that they gradually don't use it, gaining better vocal control by the third month of life. That is also helped by the physiology of the infant larynx, which at birth is located slightly farther up the throat than an adult's and only descends between the ages of three months and four years, allowing for a finer control of the tongue and mouth.

One aspect, however, seems rather clear: babies' chaotic cries are effective in getting our attention. Indeed, the human auditory system is extremely sensitive to roughness in sound and chaos. It initiates our flight or fight response—or at least evokes an emotional response.[13] Composers know this very well and use roughness of sound in horror films. These can be "scream-like" sounds—for example, the music accompanying the infamous shower murder scene in Alfred Hitchcock's film *Psycho*. This roughness of sound is used to manipulate reactions and create tension and horror, and there is certainly a lot that is horrifying in an infant's cry. "If a baby produced a beautiful song, the mother or father would say, 'They are fine; leave them to it,'" Titze said. "It wouldn't be an effective means to get your attention."

Crying, it seems, evolved to have a specific impact on listeners, and the roughness of a baby's cry makes it almost impossible to resist, as anyone who has ever tried to go to sleep while a baby cries can attest to. By driving their vocal cords to such extremes to produce these chaotic phonation regimes, newborns manage to instantly get the attention of anyone within earshot. So, the next time your newborn lets off an almighty ear-piercing attention-grabbing cry, just think that could be the bifurcation and chaos talking. Such cries will eventually make way for language (although will not disappear altogether), which will bring its own unique challenges.* And when it comes to protecting their vocal cords from damage, newborns have more in common with adult lions and tigers than they do with adult humans. So, when you get up in the middle of the night to calm that hungry baby's meltdown, hopefully you can raise a sleep-deprived smile at the similarities between your little one's complex phonation and the cartoon monkey or lion adorning their pajamas.

* It has been suggested that cries mark the very beginning of language development—more on that in chapter fourteen.

11

FEEDING FRENZY

Have you ever felt a hypnic jerk? While it sounds rather unpleasant, the chances are that you probably have, and it did not hurt. This is a sudden movement of a part of the body—usually the legs—when drifting off to sleep and, if paired with a dream, can feel like you have moved or are suddenly falling.

Newborns are born with their own innate reaction to the feeling of falling, which is known as the Moro reflex. It results in their arms abruptly flinging wide open when feeling unsupported and is usually followed by crying, possibly of the chaotic type as we have just seen. Despite your best intentions at trying to put the baby down safely in the cot, there will be moments when you do not quite do it gently enough, resulting in their arms shooting outward, startled eyes, and then some crying—maybe of the chaotic type—to deal with.

The first reflex you will likely see in action—unless the midwife drops the baby while trying to weigh the newborn—is the rooting reflex. This is when a newborn turns its head toward anything that strokes its cheek or mouth. As soon as the baby is born, that object is often the nipple to breastfeed or the midwife's finger to help kick-start the process. It can be quite amusing to see an infant frantically search for a nipple, mouth open wide, like a fish out of water.

When a baby successfully latches on, a subatmospheric pressure, or vacuum, is created that draws and keeps the nipple and some of the surrounding areola (the darker area around the nipple) in the baby's mouth. The infant then instinctively sucks on anything that touches the roof of its mouth, known as the sucking reflex. This is one of the first reflexes to appear, as early as around the twelfth week of pregnancy, with fetuses practicing it in utero with their fingers or toes. All these infant reflexes tend to disappear over time, even the sucking reflex, which comes with a use-it-or-lose-it tag (although not the ability to suck, of course). A study in 1948 that involved exclusively feeding infants with a spoon (which would likely not be allowed today due to ethical concerns) found that they lost the sucking reflex after only one week.[1]

The evolutionary twist is that once the infant starts feeding, the baby kick-starts the production of two hormones in the mother's own body. As the baby sucks, it stimulates the nerves running alongside the nipple, which promote the brain to release oxytocin that prompts muscle cells around the milk-producing cells to contract, squeezing milk into the ducts, which also dilate to help milk flow. This is known as the "let down reflex." Sucking also promotes the release of the hormone prolactin from the pituitary gland in the mother's brain. This hormone stimulates the breast to produce more milk. So, the baby is not only sucking the milk but also promoting the release and production of more of it from the mother. If the act of breastfeeding did not look so cute, you would call it parasitic.

The World Health Organization recommends infants should be exclusively breastfed for the first six months of life.[2] The decision to breastfeed is a personal one and dependent on many factors. I am not going to wax lyrical about something I have never done. All I have managed to do is gently warm up bottled milk and try to feed it to my sons without getting milk regurgitated all over my clothes, which unfortunately sometimes happened.

But it is hard to argue from a scientific perspective against the benefits for both babies and mothers. For instance, breast milk contains immune cells that help protect both mother and baby from infection. There is also evidence that breastfeeding reduces the mother's risk of breast cancer[3] as well as type 2 diabetes later in life for those who suffered gestational diabetes.[4] For babies, breastfeeding helps to prepare their facial muscles for chewing, and later, talking. But the biggest positive of breastfeeding is due to the amount of "goodness" in breast milk, such as nutrients and antibodies that are vital for the baby's growth and development.*

Colostrum, the first milk product produced by the breast, is generally available for the first sixty hours after birth. Some societies at one time thought colostrum was poisonous, and so it was not given to babies.[5] But it is now accepted to be incredibly beneficial to infants and is known widely as "liquid gold." Despite that, only a small volume, some 30 milliliters (about 1 fluid ounce), is ingested by the baby, and it soon makes way for milk. Full milk production is usually reached by the second week of lactation. The production range can vary widely, although it is around 800 milliliters (28 fluid ounces) per day for exclusively breastfed infants.[6] A study in 2016 of over one hundred women found that a third of breastfeeding mothers had reached this production capacity by day twelve following delivery, while two-thirds of mothers had done so within four weeks.[7] From then on, production stays roughly constant for the first six months of life, after which weaning tends to kick in and milk volume drops.†

Breast milk is about 87 percent water, with the rest being lactose (7 percent), fat (3.8 percent), and protein (1 percent) as well as vitamins, minerals, hormones, and antibodies. The energy from the milk is contained in the fat, protein, and lactose, with the composition of the milk changing as the infant feeds. The "foremilk" is expressed first, which is watery and quenches the baby's thirst, while the "hindmilk" is creamier and carries more fat content.[8] Fluid-dynamic simulations of milk flow through a realistic breast model by

---

* The only nutrients important for supporting normal growth in a baby that human milk don't contain are vitamins D and K. The latter can be given as an injection following birth while the former is usually administered via a supplement.

† The production of milk places an energy demand on the body similar to that of the brain.

researchers in Australia and the United States in 2017 showed that almost 50 percent of the baby's milk intake happens during the first two minutes with almost all, around 90 percent, during the first four minutes. These percentages have been backed up by clinical studies.[9]

Apart from all this nutritional goodness, human milk contains an element that had long baffled scientists: complex sugars called human milk oligosaccharides, or HMOs, which make up around 10 percent of the breast milk when the water content is removed. There are over two hundred different types of HMOs in breast milk—and their concentration is double in colostrum compared to the milk produced in the weeks following.[10] The intriguing thing is that babies cannot digest HMOs, so they pass undigested into the lower intestinal tract.* HMOs, however, are efficiently consumed by a particular microbe in a newborn's gut called *Bifidobacterium longum infantis*, or *B. infantis*. As the microbe digests the HMOs, it releases fatty acids that can be used by the infant's gut cells. *B. infantis* also promotes anti-inflammatory molecules to enhance the immune system.

HMOs also ward off pathogens in the gut, such as *Salmonella*, *Listeria*, and *Campylobacter*—the most common cause of bacterial diarrhea—by acting as a decoy so that the nasty bugs stick to them rather than the baby's own cells.[11] HMOs could, therefore, help explain why breastfed babies suffer fewer gut infections than bottle-fed babies. The ability to replicate the richness of HMOs is what is driving the development of new infant formulas and other technologies. Some manufacturers claim to have artificially manufactured specific HMOs and included them in "advanced formula," but replicating over two hundred types of HMOs seems implausible. Not only that, but breast milk has suspected "personalization" properties such as the milk changing composition during a baby's growth spurt to include more fat content or featuring

---

* HMOs are also present in maternal urine and blood during pregnancy as well as the amniotic fluid itself, meaning they could be providing benefits before the baby is born such as lung or brain development, although more research is needed before we know for sure.

antibodies when the baby is sick. And researchers are constantly discovering new aspects about breast milk. In 2021, for example, researchers found betaine—a type of amino acid also found in whole-grain foods—in breast milk. This appears to play a role in maintaining the healthy growth of a baby by fostering the development of beneficial bacteria in the newborn gut.[12]

While formula manufacturers are taking small steps to replicate breast milk, some companies are taking a more fundamental route, which includes inducing human mammary cells in a bioreactor to lactate. This would be aimed at women who cannot breastfeed but want to give their babies the closest alternative. While such lab-produced milk will not completely match breast milk, the technology could produce thousands of components, such as proteins, fats, and the whole spectrum of HMOs. Research is still ongoing and currently only small amounts of milk can be formulated. But like the lab-grown meat industry, lab-made breast milk could be the wave of the future.

Wanting to breastfeed is one thing but managing to do it successfully is quite another. According to a 2013 survey of 418 new mothers conducted by the UC Davis Medical Center in California, over 90 percent of new mothers had problems with breastfeeding. Half had trouble getting their baby to latch on or had other feeding issues, such as nipple confusion, while 40 percent said they were not sure whether they were producing enough milk.[13] Some of those concerns arise because lactation sits awkwardly between obstetrics, pediatrics, and general family health. Historically, obstetricians would focus on the pregnancy while pediatricians would focus on the baby. But the interaction between mother and baby was often neglected, with no expert available on breastfeeding. That has changed to some extent recently given that, in the United Kingdom for example, the midwife is the one who takes care of lactation. However, midwives in some countries like the Netherlands do not see breastfeeding as their responsibility. This has fostered a new industry—"lactation consultants." But whether they have really plugged the care gap is debatable.[14]

With our firstborn, Henry, we had breastfeeding issues that were never fully resolved. He seemed to latch on well and was sucking away for long durations. But at successive midwife appointments in the first weeks after birth, he was found to be losing weight. While it is normal for breastfed newborns to lose about 10 percent of their weight after birth due to the loss of body fluid, that weight loss is generally made up within the first two weeks. When he continued to lose weight, the midwife referred us to the local children's hospital, where his feeding was observed. Everything seemed fine, and he was obviously getting something, but presumably not quite enough. (He was never weighed before and after a feed, which could have given an indication of how much he was getting.) It was then advised that we supplement with formula. Of course, this did the trick, and he immediately began gaining weight, but it was frustrating not to know what the reason was for the weight loss. Was it a lack of milk production, not a strong enough latch-on, or perhaps a problem with the actual feeding technique?

While there are large research programs to examine the composition of breast milk, for all the benefits of breastfeeding, there is still a lot we do not know. This includes the seemingly simple issue of how an infant manages to extract milk from the breast in the first place—a question that has vexed scientists as far back as the late nineteenth century. It is obvious that an infant uses vacuum suction to get the breast in the mouth. The only way to get a well-latched-on baby to come off the breast is to break the seal by putting your little finger in the side of its mouth.

But the question is whether feeding itself is the result of a change in vacuum pressure in the infant's mouth (i.e., cycling between a negative pressure greater than the baseline pressure to get the nipple in the mouth). Or it could be that the feeding process is a wavelike, or peristaltic, motion of the tongue as it compresses the nipple from front to back to "strip" the nipple of milk, somewhat like a dairy farmer hand-milking a cow's teats. Both options sound plausible. After all, modern breast pumps use a cycling of subatmospheric pressures to successfully remove milk from the breast without any peristaltic movement on the nipple at all. On the other hand, literally, it is possible to hand-express the breast to produce milk without the need for vacuum pressure. Thanks to modeling and image analysis, some of these aspects are now being elucidated and, in the process, helping scientists get closer to answering that century-old question of whether breastfeeding can be defined as "sucking" or "suckling."

Feeding babies milk is unique to mammals, and the delivery mechanism for all this goodness is, of course, the breast. The anatomy of the breast was first documented in 1840 by the British surgeon Sir Astley Paston Cooper. Regarded as one of the world's greatest physicians, Cooper likely obtained breast samples from the bodies of cadavers that were often provided by "resurrection men" who secretly went about removing bodies from burial sites.[15] Cooper's careful dissection and detailed anatomical drawings set the foundation of our understanding of the lactating breast—one that is still used in textbooks today.[16] Via the examinations, Cooper found that the breasts are made up of fatty, fibrous, and glandular tissues, embedded in which are fifteen to twenty glands called lobes. Each of these lobes contains numerous smaller sacs called lobules that are bunched together like grapes. There are around ten thousand lobules in a breast, with the milk produced by them traveling via a single thin tube, or duct, to the nipple. Despite all the ducts leading to the nipple, it is thought that only around a third of these are open at the nipple tip.

In the late 1800s, scientists thought that the answer to how a newborn feeds from the breast would require simply measuring the pressure an infant could produce. Yet, accurately doing so would be experimentally challenging, and much of the early work involved building, in a simple sense, artificial breasts. This consisted of an artificial teat that had two tubes attached to it, one that led to a reservoir of milk that the baby drank while the other was connected to a "manometer," an instrument still used today to measure pressure. The manometer consists of a U-shaped column that contains a fluid—usually water or mercury—of known weight, which sits at the bottom of the U-bend, with the liquid initially measuring the same "height" on both sides of the U.

Fig 1 | *Breast ducts as imaged by Sir Astley Paston Cooper in 1840. Jefferson Digital Commons.*

When a suction pressure is applied to one side (with the other end closed off), the liquid rises up on that side of the tube while the other side drops. How much this "height" changes relates to a pressure—the higher the change in height, the larger the pressure.

In the 1890s, the Austrian-Jewish physician Samuel Siegfried Karl von Basch, who is best known for inventing the blood-pressure meter, used the artificial-teat device to find that an infant could generate sucking pressures of about -10 mmHg.* Basch then put a pump on a breast and noted that he could not extract milk until he reached -40 mmHg.[17] Given this discrepancy, he concluded that infants do not have enough sucking power to extract milk from the breast, and so they must be stripping the teat instead.[18]

The Austrian pediatrician Meinhard von Pfaundler, however, disagreed with Basch's conclusions. Von Pfaundler thought that infants create sub-atmospheric pressures in the mouth to suck milk from the breasts. To test his theory, he had babies drink a column of milk that was connected to an artificial nipple and found that "strong" babies could suck milk up nearly 70

* Negative pressures are represented by a negative sign.

centimeters high while "weak" babies could manage only around 20 centimeters.[19] He concluded that babies generally had enough suction strength to extract milk in this way from the breast.

In 1951, the physiologist Frank Hytten from the Department of Midwifery at the University of Aberdeen in Scotland used a manometer to discover that the sucking reflex could generate a maximum suction of -50 mmHg.[20] "It is probable that higher values reported by other workers were obtained by the use of faulty apparatus," he wrote. He found he could replicate a high pressure if the infant's tongue sealed off the tube or teat, which caused a "valve-like action" that causes negative pressure to build up that exceeds the "true" oral pressure. Focusing on such measurements was all well and good, but they did not really tell you much about the mechanics of *how* infants feed.

When you watch an infant breastfeed, mesmerizing as it is, all you see from the outside is the rhythmic moving of the lower jaw up and down to compress the areola. Yet, on the inside, a lot is happening as the tongue moves up and down and back and forth to swallow the milk. The first clear look at this in action was in the late 1950s, when scientists from the United Kingdom and Denmark used X-rays to study forty-one infants breastfeeding. To take the images, the scientists coated the nipple in a mixture of lanolin, a waxy substance secreted by sheep, and barium sulphate. Barium, a white, chalky material, is a good absorber of X-rays and so shows up well on scans—a bit like calcium in bones.

The X-ray films revealed that, as the infant sucks, it takes a large amount of the areola into the mouth in addition to the nipple. This creates a "teat" that extends three times as long as the nipple, reaching at its farthest point at the junction between the hard and soft palates in the upper mouth. The film also showed, for the first time, that the tongue plays a crucial role in breastfeeding. As the jaw rises to compress the teat against the hard palate at the top of the mouth, the tongue also compresses the nipple, but it does so progressively from the front to the back like a wave. This led the team to conclude that the vacuum holds the teat in place while the infant performs a milking, or stripping movement, to obtain the milk from the breast.[21] The work lay low for some time but eventually received considerable attention due to the amount of information it unveiled about infant sucking. It remains a one-of-a-kind study given the ethical concerns today of using ionizing radiation for such research.

Further evidence of a stripping motion emerged in the late 1980s, thanks to an ingenious technique devised by Kazuko Eishima at Chikushi Jogakuen Junior College in Japan. She was working at Queen Charlotte's Hospital in London* on newborn reflexes, including the sucking reflex. To do so, Eishima built a device that enabled her to directly observe the inside of an infant's mouth via a camera attached to the bottom of a bottle that had a transparent artificial teat. She also used a fiberscope, made up of a bundle of optical fibers, that allowed her to film the sucking motion of the baby's mouth. Eishima studied 287 infants, recording fifty of them during both active feeding and so-called "non-nutritive sucking," which happens at the start and end of a feed and is characterized by a series of short bursts and rest periods but no removal of milk and no swallowing.[22]

Analyzing the video footage, Eishima concluded that the tongue made a peristaltic movement from front to back as an infant feeds.[23] Yet, when she used a teat with a larger hole, the baby was swamped with fluid and the wave-like movement of the tongue disappeared, perhaps due to the infant needing to breathe and swallow the milk at the same time. Intriguingly, Eishima also found that the surface of the tongue at the nipple tip has a periodic downward "dip," suggesting perhaps that infants create added suction during feeding. The study gave insight only into bottle feeding, which differs from breastfeeding, given the material difference between the nipple and an artificial teat. Nevertheless, it offered more evidence for stripping behavior.

From the early 1980s, researchers switched their focus to what real-time ultrasound could offer in the breastfeeding conundrum.[24] Ultrasound scans are often conducted with the infant held by the mother in the "cradle position," and then the probe is placed under the infant's chin in such a way that it does

* In 1988, Queen Charlotte's Hospital and Chelsea Hospital for Women merged into one site that was renamed Queen Charlotte's and Chelsea Hospital.

not impact the latch-on to the breast. Using ultrasound, researchers in the United States found that the nipple was particularly elastic, almost doubling in length. They deduced that nipple compression draws milk into the ducts, but the release of the milk is caused by a vacuum effect by "rapid enlargement of the oral cavity."[25] Despite improvements in ultrasound technology at the time, scans were still somewhat noisy, reducing image quality. After all, when an infant is breastfeeding, there is a lot going on—the mother is breathing, the infant is sucking and swallowing, and in general, just moving around. It was not until a decade later that the image resolution was sufficient to see what was going on in any meaningful detail.

Donna Geddes at the University of Western Australia has pioneered the study of breastfeeding via ultrasound and arrived at this area of research in a rather serendipitous manner. She was a clinician, carrying out ultrasound examinations on the blood flow through the breast, when in the early 2000s she attended a conference at the university and met lactation specialist Peter Hartmann. He was looking for someone to join his group and study the let-down reflex using ultrasound. Given Geddes's background, it was a perfect match for her expertise, and following a postgraduate diploma in science in lactation, she completed a PhD in his group that was partly sponsored by the Swiss company Medela, a leading supplier of breast pumps and feeding bottles.

In 2008, Geddes and colleagues performed ultrasound investigations on twenty babies who were between three and twenty-four weeks old. Along with the scans, the team also measured the pressure in the infants' mouths as they fed, which was done via a tube connected to a transducer. The tube was inserted into the mouth with the rest of it turning alongside the nipple and breast to the transducer. Similarly to Eishima's bottle study, they discovered that the part of the tongue next to the nipple tip moved slightly downward. When this happened, Geddes measured a vacuum pressure of about -150 mmHg—much stronger than the baseline level of -60 mmHg that kept the nipple stretched inside the mouth. When that dip occurred, which they dubbed an "intra-oral vacuum," the researchers also saw on the ultrasound scans that milk was ejected into the infant's mouth.[26] From this, they concluded that a critical component of breastfeeding was the tongue dip, which added additional suction. It was vacuum pressure, rather than peristaltic movement of the tongue, that allowed the infant to feed. A follow-up study in 2012 backed up the claim, finding that during non-nutritive sucking, the surface of the tongue at the nipple tip did not drop as much as compared to nutritive sucking.[27] With two studies suggesting peristalsis and another putting forward vacuum, it remained elusive which technique infants use. Perhaps an engineering approach might help?

David Elad, a bioengineer at Tel Aviv University in Israel, has spent most of his career working on problems associated with reproduction, whether it is how the blastocyst implants into the uterus, how the heart forms in morphogenesis, or the dynamics of uterine contractions. He completed his PhD in biomedical engineering in 1982 from the Technion–Israel Institute of Technology, and after a stint at Northwestern University in Illinois, returned to Israel.

Well known for his expertise in bioengineering, Elad was approached by a surgeon at Tel Aviv Sourasky Medical Center who was having difficulty analyzing some ultrasound scans that showed an infant struggling to breastfeed. The surgeon suspected that the infant had a condition known as tongue-tie, or ankyloglossia, that restricts the tongue's range of motion. This is caused when the frenulum, or the tight band of tissue that connects the tongue to the bottom of the mouth, is attached too far forward under the tongue or is too stiff. The condition affects as much as 11 percent of newborns in the United States and is often treated via an incision in the frenulum that releases the tongue.[28] Yet, the use of this surgical procedure—or rather, overuse—remains controversial.

The problem, however, in helping the surgeon analyze the ultrasound images was that Elad had no reference point to compare the images to. So, he asked if the hospital could carry out some ultrasound examinations on healthy infants breastfeeding, and he eventually received scans of nine infants.* Given

* For a selection of ultrasound movies of an infant feeding, see: www.eng.tau.ac.il/~elad /Lab/movies.html.

the amount of movement in the scans, to get an accurate measurement of what the tongue was doing for healthy infants during breastfeeding, the researchers used the top of the hard palate on the roof of the mouth as a fixed point, knowing that this area should not move in relation to the tongue and nipple during feeding. Using this fixed point in a technique called "rigid registration," they measured the outline of the tongue and nipple, as well as the soft palate of the mouth toward the back of the throat, as the infant breastfed.

They found—just as the X-ray study did in the 1950s—that when latching on, the infant uses vacuum pressure to extend the "teat" near the junction between the hard and soft palate, which is about 25 millimeters (close to an inch) from the lips of the infant. Yet, when they then looked at the movement at the front of the tongue, which lay under the nipple, they found that it moved like a solid body—there was no wavelike behavior at the front at all. Movement was basically controlled by the periodic motion of the lower jaw and did not act to strip the nipple[29] as had been presumed before in other studies (apart from the Geddes study).

The researchers did see peristaltic movement of the tongue, but this started at and beyond the tip of the nipple—so not on the nipple itself—and moved progressively toward the back of the tongue. This movement allows the infant to swallow the milk, as the baby would be unable to do so using suction alone.* The "teat," meanwhile, moved back and forth and was in step with the movement of the front of the tongue, taking about 0.6 seconds to complete one suck—so fairly rapid. Elad concluded from this analysis that infants feed by a rigid compression of the base of the nipple combined with varying vacuum pressure to promote milk flow that is then swallowed at the back with peristalsis.

Despite these findings, physiologist Michael Woolridge, who has spent his whole career working on many different aspects of infant feeding, is convinced that peristalsis plays a role at the front of the tongue. With a PhD in zoology from Oxford University in 1976 (supervised by Richard Dawkins of

---

* It is known that peristalsis is at play at the back of the tongue to aid swallowing in both babies and adults.

The Selfish Gene and *The God Delusion* fame), he initially began working on the nutritional aspects of breast milk but became increasingly interested in how an infant takes milk from the breast. "The practical aspects of breast-feeding have never been quantified, in contrast to the massive attention to the constituents of breast milk," he told me during a phone conversation.

Woolridge, who began carrying out ultrasounds on feeding infants in 1990,[30] insists that despite what other studies say about how critical vacuum is, it is impossible to ignore peristalsis as the main mechanism in milk extraction. Looking at his own research, along with previous work that has shown peristaltic action, he says it is well known from a "neurological perspective" that the sucking reflex generates a peristaltic wave starting at the front of the tongue.

In 2011, Woolridge and Gianluca Monaci, who worked in signal and imaging processing at Philips Research in the Netherlands, went about trying to prove this by carrying out ultrasound recordings of twenty-nine babies feeding, in which they compared the amount of movement at the front and back of the tongue. They found that peristaltic movements were central about 50 percent of the time, with intra-oral vacuum dominating for 20 percent of the time (the remainder being a mix of peristaltic and vacuum movements).[31] The duo also designed a program that automatically tracks the surface of the tongue—both front and back—as it moved, discovering that the tongue dipped at the nipple tip to create a "vacuum pocket," exactly as Eishima and Geddes had discovered (although it was not present in Elad's study). Woolridge admits the vacuum pocket shows that cycling vacuum pressure is important, but he insists it is not the dominant feature.

In Woolridge's view, then, babies both suckle, or strip, *and* suck from the breast to retrieve milk. As the infant chomps on the breast from the jaw—applying a pressure to the breast tissue—this promotes milk flow into the milk ducts toward the nipple. Then peristaltic tongue movements compress the nipple at the base to squeeze milk to the tip and out. Woolridge thinks that the extra vacuum provided by the tongue dip at the nipple tip acts to extend the duration of the milk flow, boosting the amount of milk. According to Woolridge, infants tend to tongue dip every second suck and always after a swallow of milk, with the tongue dip enhancing the ability of the next suck to extract milk. "At the end of the day, it is the variability of peristalsis and vacuum when breastfeeding that shows how adaptive babies are," Woolridge said. While some would disagree with Woolridge's analysis, it is true that an infant cannot breastfeed without creating subatmospheric pressure in the oral cavity *and* performing a peristaltic movement at the *back* of the tongue to swallow the milk.

One issue with ultrasound is that researchers only get a 2D view. A fuller understanding of the tongue's movement requires a 3D view. After all, the tongue is not a flat paddle. Despite that, however, the research shows one important thing: to optimize breastfeeding, the infant needs to take as big a mouthful of the breast as possible to ensure the "teat" extends farther into the mouth and give the chance for the tongue to do all the action and maximize milk flow. And while it remains inconclusive how infants feed, perhaps the bigger aspect of studying the mechanics of breastfeeding lies in the potential diagnosis of conditions that affect breastfeeding as well as improvements in technology for better bottle teats and pumps.

Elad and colleagues are using their analysis techniques based on infant sucking to investigate ways to diagnose and treat breastfeeding issues. In 2021, they studied infants that had either tongue-tie, dysphagia (a condition that results in swallowing difficulties), or lip-tie, which is when a piece of tissue behind the baby's upper lip is too short and tight and limits the upper lip's movement.

In infants with dysphagia, the team found that the back of the tongue, as well as the front, did not undergo peristalsis at all, further cementing the tongue's critical role in swallowing the milk. In infants with tongue-tie, they discovered that the tongue underwent very chaotic movement during feeding without the smooth periodic feeding shown in a healthy infant. However, when Elad did the same analysis after the infant underwent the small surgical procedure, the movement became like that of a baby with no tongue-tie.[32] A similar effect could be seen in an infant with lip-tie before and after surgery.

"The study provides an objective method to explain the efficiency, or deficiency, of breastfeeding," Elad noted.

As for Geddes, she agrees that further developments in ultrasound and the analysis of breastfeeding mechanics could eventually lead to new diagnostic tools for women who struggle to breastfeed. But perhaps before these techniques can become widespread, a lot more support needs to be forthcoming. As Woolridge admits, while it is relatively easy to find funders who will support studies of breast milk, doing so for other important aspects of breastfeeding, such as the mechanics, is much more difficult. "People always talk about how important breastfeeding is for infants," Elad said. "But when it comes to understanding the mechanics, it is an area that for too long has been underfunded and neglected."

THIRD INTERLUDE

BREAST PUMPS AND BABY BOTTLES

You might think that breast pumps are relatively modern inventions, but they have been around since the days of the ancient Romans and Greeks[1]—albeit not in the guise that we know them now. The first recorded device was built in 1854 by inventor Orwell H. Needham from New York who patented a hand-operated mechanical device simply called a "breast pump." It was based on a suction bulb, which was connected via a tube to a glass cup that was placed around the breast. Further innovations followed, such as removable milk-collection bottles in 1874. These early devices were based on the mechanical designs used to milk cows in the dairy industry. It was not until the late nineteenth century that devices began to mimic the way human infants feed. Inventor Joseph Hoover from Iowa patented a device that generated a "pulsating movement" via a spring, which relieved the breast from the constant pull that occurred with previous contraptions. Unfortunately, many of these hand-operated devices were not able to remove a large enough milk supply from the breast and were slow. They were principally operated by squeezing a bulb and waiting for it to repressure before squeezing it again.

A breakthrough came in 1928 when inventor Woodard Colby designed a device that could keep a constant vacuum, which enabled a suck-and-release cycle to retrieve milk, similar in some sense to how modern breast pumps work. This allowed many more cycles of pumping per minute than previously possible.

Breast pumps at the time were still medical devices, used to treat conditions such as inverted nipples or to help preterm babies or infants who could not produce enough vacuum pressure in their mouth to feed. Comfort, however, was often an afterthought (as some would say is still the case today) or at worst not thought about at all. Improvements arrived in 1942 with the first hospital-grade electric breast pump. Designed by Swedish engineer Einar Egnell, the pump used a maximum suction pressure of -200 mmHg.

The 1940s and beyond saw a surge of companies interested in the breast pump market. Ameda, founded in 1942, acquired Egnell's pump to sell to

Fig 1 | *Breast Pump, London, England, 1870–1901. Science Museum London.*

hospitals. The Swiss firm Medela was founded in 1961, and in 1980 created a mobile pump that could be wheeled around a hospital. By the late 1990s and early 2000s, both companies had released the first home-use breast pumps, with many other companies since following suit with their own innovations. Despite the flexibility that breast pumps brought—allowing women to exclusively breastfeed and go to work, for example—there is still a lot of room for improvement. They are loud, restrictive, expensive, and, worst of all, they can be painful. I can still to this day hear the pulsating sound of the breast pump my wife used and how it visibly moved the whole breast in a sucking manner—completely different from what is observed when a baby feeds. "Modern breast pumps are like vacuum cleaners," said Israeli bioengineer David Elad, who added that pumps operate at too high a pressure—around -150 and -200 mmHg—and work on a tug-and-release action, which can be painful.

Initiatives have been created to tackle some of those issues. The Media Lab at the Massachusetts Institute of Technology, for instance, has held regular "hackathon" events in the past decade to bring together hundreds of parents, designers, engineers, midwives, and breast pump manufacturers to discuss problems and test solutions.[2] It has resulted in several innovations, such as the Mighty Mom Hush-a-Pump case that can reduce the noise of a typical modern breast pump by about 50 percent.

The reason why breast pumps operate at such high pressures is because it is effective. After all, no one wants to spend hundreds of dollars on a pump, get it home, and then find it does not work or takes hours to get just a few milliliters of milk. A 2008 study by Donna Geddes at the University of Western Australia and her colleagues showed that the optimal way to extract milk is to use the "maximum comfortable vacuum." This is defined as 10 mmHg less than the pressure at which you personally find it painful to pump.[3]

In the study of twenty-one women, half could tolerate a pressure higher than -200 mmHg, and expressing for fifteen minutes around this value resulted in more milk at quicker rates than using lower vacuum levels such as -125 or -80 mmHg. The women were unable to extract milk at all with a vacuum pressure below -80 mmHg, and so the authors concluded that to maximize the amount of milk and minimize the duration of pumping, mothers should use their own maximum comfortable vacuum, however high that may be.

Once milk has been extracted from the breast it needs to be stored and delivered to the baby, which is usually done via a bottle and artificial teat. In a similar manner to breast pumps, one might think that bottles are a modern invention, but using a device to deliver milk to a baby has been in use since ancient times. The first clear evidence of feeding vessels dates to around 2000 BC. In 2019, researchers analyzed the deposits in several small clay containers taken from the graves of Bronze Age newborn infants in Bavaria, Germany.[4] It was long thought that these vessels, which look somewhat like a genie's bottle, were used for other purposes than infant feeding. But chemical analysis showed that signatures of fatty acids from animal milk was present in the containers' residue.

Since ancient times, infants that were not breastfed by their birth mother (perhaps due to maternal death) would have been wet nursed, in which a woman breastfeeds another's child. This was the safest and most common alternative to breastfeeding by the birth mother. Over time, the increasing negative stigma associated with wet nursing, combined with the availability of animal's milk and the bottle, gradually led to feeding substitutions.[5] Sometimes this simply involved infants taking milk directly from the animal's udder, the most famous depiction of this perhaps being the bronze sculpture *La Lupa Capitolina* (The Capitoline Wolf) showing the twin brothers Romulus and Remus, whose story tells the founding of Rome, guzzling milk directly from a wolf, saved from oblivion by an animal wet nurse.

Of course, it wasn't possible for all children not breastfed to have their own personal round-the-clock access to a cow, goat, sheep, or even donkey.[6] There was also the problem of the milk going foul just hours after being extracted. So, feeding indirectly from animals or from expressed breast milk required a suitable container from which to suck. A common type of feeding bottle during the Middle Ages was a cow's horn, which had a small hole at the tip that would transform it into a feeder.[7] By the seventeenth century, infant feeding devices in Europe included ones made from wood, pewter, glass, silver, and ceramics. They mostly resembled teapots—a vessel with a spout and a handle on the opposing side. Cloth or sponges were also placed over the holes so that the milk would filter through.

Innovations soon followed, including in 1700 when Hugh Smith, a physician working at Middlesex Hospital in London, invented the "Bubby pot."[8] Made from pewter, it resembled a small coffeepot with the spout emerging from near the base. The end of the spout had a roundish knob that Smith described as being "in appearance like a small heart." The end of the spout included three or four small holes, and a piece of rag was tied loosely over it to strain the milk. Smith observed that infants who used it didn't get overwhelmed by milk flowing out of the spout, allowing feeding to be a smoother process.

Despite successes in design, a huge issue at the time was keeping these devices clean as the innermost areas would often teem with bacteria. When

Fig 2 | *Bubby pot for infant feeding, England, 1770–1835. Science Museum London.*

combined with a lack of proper milk storage and sterilization, it has been estimated that in the nineteenth century a third of infants that were artificially fed during the first year of life died.[9] Strides were made to make feeding from the bottle safer and easier. Glass bottles were widely introduced, and later designs included heat-resistant glass. Rubber teats were also created, and while they initially had a repulsive odor and taste, developments at the start of the twentieth century improved their use.

Today, we have plastic bottles as well as silicone nipple-like teats that offer a range of different flow rates (although our children still managed to drench themselves in milk from time to time). And innovations are still being made. Geddes says that a knowledge of the mechanics of breastfeeding (see previous chapter) is improving things. When infants take the bottle, they tend to feed how they would do so on the breast, but with more emphasis on producing a greater vacuum suction. The mismatch in the material characteristics between the nipple and artificial teat can cause the tongue to produce a more chaotic movement. Bottle teats tend to be more easily compressed than nipples and do not reshape themselves to fit into the infant's mouth, which can be problematic as babies have to keep up with the continuous milk flow and breathing at the same time, resulting in the ejection of milk out of the mouth or coughing and choking.

Developers of modern baby bottles are now trying to address some of those issues by letting the baby control the flow of milk. Some are engineered so that milk flow is regulated by the baby's sucking and milk flows out from the bottle in specific ways. This means that solely compressing the teat does not extract milk, with the baby having to feed via applying a vacuum suction as it would do on the breast.

The aim is to give the infant the nearest possible experience to breastfeeding. But this is not just a modern ambition. As Smith wrote when he designed his Bubby pot back in the 1700s: "The child is equally satisfied with it as with the breast: it never wets him in the least: he is obliged to labour for every drop he receives . . . and it saves a deal of trouble in the feeding of an infant."[10]

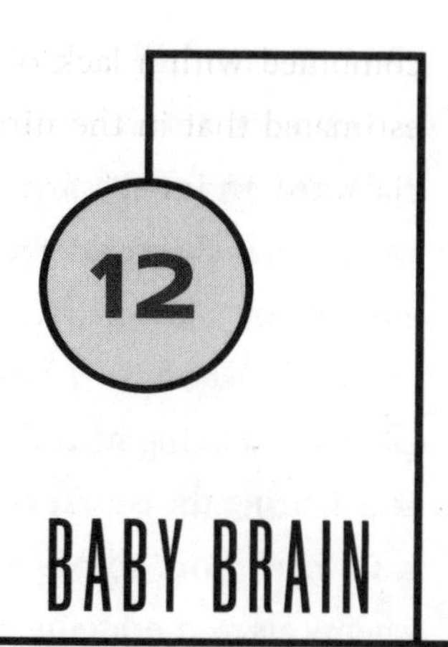

BABY BRAIN

There is no single bigger issue for parents than sleep—both yours and your newborn's. How many hours did I get last night? How many naps should infants have during the day, and when do they start dropping naps? When can I finally expect my child to sleep through the night? (Not soon enough.)

The saying "sleeps like a baby" exists for a reason, and it is true that newborns sleep a lot. In the first two weeks of life, they sleep for around sixteen to eighteen hours every day on average. The "on average" range is rather large, with some babies blissfully sleeping nineteen hours a day while others, horrifyingly, get by with only nine hours.[1] In the beginning, knowing that babies sleep a great deal can give new parents a false sense of security, but by the end of the first month, the total amount of sleep has already dropped by two hours, on average.

The problem is that while babies sleep a lot, they also wake up a lot, and that includes during the night. The main reason is that a newborn's stomach is rather tiny, so each sleep session can last anything between thirty minutes to four hours before they wake up with a hungry chaotic cry. Babies also do not have a circadian rhythm, so their internal body clocks are not yet synchronized to the twenty-four-hour day. This sense of day and night begins to develop when newborns are between twelve and twenty weeks old, and there is not much you can do to speed things along, unfortunately.[2] The same goes for when an infant starts to sleep through the night. It is thought that a hint of a regular sleep pattern only begins to emerge at eight months, taking a year to fully, erm, bed in.

Scientists have proposed many reasons why we sleep—the two most prominent being neural repair and neural reorganization that leads to learning. During the course of the time when we are awake, the brain suffers "wear and tear" from things such as blood flow and the production of harmful chemicals and proteins. Sleep helps to clean this up. A good night's sleep for an adult—something that new parents can only dream about, when they get sleep, that is—is seven to nine hours.

Adults sleep in four stages that take around ninety minutes for a full cycle, meaning five to six sleep cycles are completed, hopefully, each night. The first stage is a light sleep, a transition between wakefulness and sleep. This is then followed by the second stage in which body temperature lowers and brain activity begins to slow down, making it more difficult to wake up. The third stage is the deepest sleep, when the muscles relax and blood pressure and breathing rates drop before one enters the final stage, called rapid-eye movement (REM) sleep, in which the heart rate, breathing, and eye movement all speed up. Adults spend about 20 percent of sleep in the REM stage.

Newborn sleep is different. Adult-like classification of sleep cycles is not possible until the baby is around two to three months old.[3] Sleep cycles in newborns are shorter than adults—around fifty minutes—and they tend to only have two main stages of sleep. One is "quiet sleep" when the baby appears more restful with slower and more rhythmic breathing and is harder to wake up. This is the time when parents coo over their delightful creation despite the increasingly dark circles that appear under their own eyes. Sleep changes a lot during the first five weeks, with non-REM sleep increasing from about 10 percent at two weeks to 20 percent at five weeks.

Newborns spend most of the time—between 50 to 75 percent—in "active sleep," which is basically a baby version of REM sleep.[4] This sleep state is characterized by fluttering eyelids; rapid, irregular breathing; body movements; and the occasional grunt. All this action can seem like your little one is about to wake up, but the baby is still asleep. And, let's be reminded—everyone has heard the saying, "Don't wake a sleeping baby." The role of this baby version of REM sleep has been a bit of a mystery to scientists, particularly as it is so high in infants. The main explanation is that this active sleep supports their fast-developing brains, especially in the first six months of life.

The amount of REM sleep a person needs gradually decreases throughout life, so that people over fifty years old, for example, spend only 15 percent of sleep in REM. In 2020, an interdisciplinary team of scientists in the United States looked at sixty previous sleep studies that examined total sleep time—time in REM sleep, brain size, and body size for children from birth to fifteen years old.[5] Analyzing all this data, they discovered that REM sleep decreases with increasing brain size or age, but at exactly 2.4 years something remarkable happened. They found a sharp transition in the primary purpose of sleep, as it moved from predominantly focused on neural reorganization to concentrating on neural repair, which is then maintained throughout life. What is surprising is that this is not simply a gradual adjustment from organization to repair, but an abrupt change.

Sleep during the night for both babies and adults is much more complicated than just entering deeper sleep before coming back out again, like perfect steps on a bar chart. There is not only flip-flopping back and forth between cycle stages but also spikes of brain activity that briefly jolt you out of deep sleep while keeping you still asleep, or from light sleep to awake. This tends to happen when transitioning between different sleep stages, and such brief awakenings, or arousals, can occur in adults, children, and newborns and at any time of night.

A statistical analysis of these transitions shows that there is no typical length for arousals in adults—it can be anything from a few seconds to minutes—but there is a characteristic duration between them: twenty-two minutes, giving between ten to fifteen arousals per night.[6] This same kind of behavior has also been seen in other animals, with cats having an eleven-minute duration between arousals while mice have six minutes.[7] The physics of these events follows a behavior called self-organized criticality. A familiar example of this is a sandpile in which sand grains are slowly sprinkled at the same point to create the pile. After a while, the amount of sand forming the pile will become too much and it will set off avalanches. In a similar vein, brief arousals can be classified as "avalanches" in the wake-sleep mechanism of the brain.

Sleep scientists have been stumped as to what triggers these nocturnal disruptions, but physicist Ronny Bartsch at Bar-Ilan University in Israel found them intriguing from a physics perspective. Bartsch first became interested in sleep research after hearing a talk in 2003 by the physicist Thomas Penzel, head of sleep research at the Interdisciplinary Center for Sleep Medicine in Berlin. At the time, Bartsch was studying for his master's degree at the University of Konstanz in Germany on the physics of cardiac dynamics, but he was so inspired by Penzel's lecture that he switched research fields. He pursued a PhD at Bar-Ilan University where he used methods in statistical physics to analyze sleep patterns.

After his PhD studies were over, Bartsch headed to Harvard Medical School in Boston in 2008. The next few years were a particularly exciting time for Bartsch, not just professionally, but personally, too. In 2012, he and his wife were expecting their first child, a girl. But she was born prematurely, at just twenty-four weeks' gestation, and immediately transferred into a neonatal intensive care unit (NICU) at Harvard Medical School where she spent sixteen weeks with round-the-clock care. Despite the challenges of being born at just 650 grams (less than 1.5 pounds), she did incredibly well, leaving the NICU (exactly on her due date) at a healthy 3.7 kilograms (about 8 pounds)—as much as a full-term baby. "It was an incredibly long and tough time," Bartsch told me over a video conversation. "The only solace being that she was being looked after at one of the best places in the world."

When Bartsch's daughter was in the NICU, he began to read about sudden infant death syndrome (SIDS) and infant sleep and became intrigued about how different newborn sleep is from adult sleep. "Much of what happened during those early days with my daughter really triggered my interest in infant sleep," adds Bartsch. Given that preterm babies are at a higher risk of SIDS, Bartsch became much more aware of its dangers. SIDS usually happens when the baby is asleep, but not always, and there are several risk factors that are associated with SIDS, such as whether the parents smoke, co-sleep with their baby, put the baby to sleep on its tummy, or wrap the baby in too many blankets.[8] What makes it so disturbing is that deaths tend to be unexpected and unexplained and can occur in otherwise healthy babies, both full-term and preterm babies.

SIDS is thankfully rare. In the United States in 1999, for example, the number of deaths associated with SIDS stood at 130 per every 100,000 live births. But that figure today is just 35 per every 100,000 live births.[9] One reason behind this drop is awareness and education. New parents are routinely told to put the baby to sleep on its back, avoid smoking indoors, and make sure the baby's feet are at the bottom of the cot so it cannot wriggle down underneath sheets, as well as make sure the room is not too hot. Newborn babies make up around 10 percent of all SIDS deaths, with the condition peaking at 30 percent for two-month-olds before rapidly decreasing, so that at six months old only around 2 percent of babies who die of SIDS are this age. At nine months, the number of deaths is almost zero,[10] and this risk drop is thought to be because the baby can roll around or kick bedding off, if they end up with a blanket on their face, for example. With the days in the NICU behind him, Bartsch began to apply statistical physics to a neuronal model of sleep arousals, and in the process, he discovered an intriguing possible link with SIDS.

Sleep, whether in newborns, infants, or adults, is due to the dance of signals that occur in neurons, or nerve cells, in the brain. Neurons are the building blocks for how the brain operates. They are electrically excitable cells located in the brain and spinal cord and come in many different shapes and sizes. Yet, what they all have in common is a strange, alien-like appearance. Neurons receive their signals via dendrites, long branched projections from the cell body. If the signal crosses a certain threshold, then the neuron sends a pulse of activity down long, thin axons that act like biological cables. At the other end of the neuron is the axon terminal, with a small gap called a synapse located between the terminal and a neighboring dendrite of another neuron. The brain contains around eighty-six billion neurons, and as every neuron has seven thousand connections, it is estimated that an adult brain has about 250 trillion synapses.

Neurons are excitable thanks to a voltage difference that exists across the cell membrane caused by sodium and potassium ions as they rush in and out of neurons, as described by the Hodgkin–Huxley equation, which we came across in chapter six. When a neuron has been sufficiently stimulated by its neighbors, ion gates that were locking out sodium ions suddenly open, triggering an action potential. A massive influx of these charged particles ensues, causing the interior voltage to spike from a resting potential of -70 millivolts to +50 millivolts—a jump of about 120 millivolts, or 0.12 volts. While electrical signals in a circuit can travel near the speed of light,* the signals in a neuron travel around 120 meters per second (about 270 miles per hour). This is because the electrical signal is converted into a chemical one and back again, and this conversion slows things down. Despite that, signals can still be transmitted in a few thousandths of a second so that it quickly propagates to other connected neurons.

Neurons can receive thousands of signal inputs, and these can be excitatory or inhibitory. Excitatory inputs boost or "add" to the neuron's overall signal while inhibitory inputs decrease or "subtract" the signal in a process called synaptic integration. Sleep is the interplay of all this neuronal activity, predominantly in two areas of the brain: the brain stem at the base of the brain and the hypothalamus, a peanut-sized structure deep in the organ. These parts have wake-promoting groups of neurons and a sleep-producing group, which result in a flip-flop circuit that provides sleep-wake control.[11] The neurocircuitry of how it all works is complex and not fully understood,

* The speed of light in a vacuum is 299,792,458 meters per second.

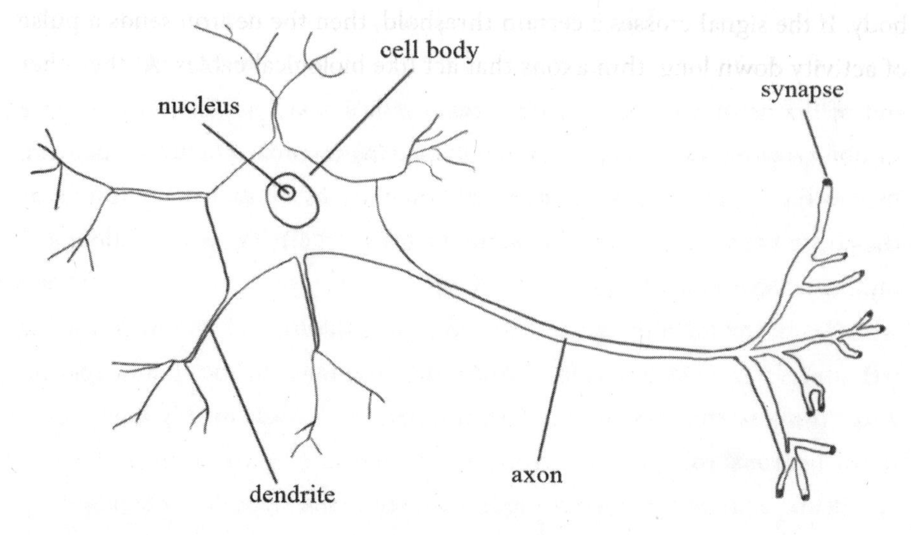

Fig 1 | *Schematic of a neuron.*

but simply put, wake-promoting neurons tend to be under circadian control and are active during the day, while it is thought that the sleep-inducing neurons increase in activity the longer one is awake. The result of all this interplay is that when one system inhibits the other, there can be a switch between wakefulness and sleep. To appreciate the power of the battle that occurs in your brain, all you need is a wide-awake newborn in the middle of the night.

Key to the communication between neurons are neurotransmitters—molecules that are often referred to as the body's chemical messengers. When an action potential travels down the axon, it results in voltage gates opening at the axon terminal that release neurotransmitters. This then binds onto the neighboring dendrites and sets off positive ions to flow—what is known as a synaptic potential—in the connecting neuron. Neurotransmitters are also generally excitatory or inhibitory. The latter produces greater negative voltage, which makes the neuron less likely to fire an action potential, while excitatory neurotransmitters repolarize the cell making it more likely to fire, and once this summation of neurotransmitters reaches a certain threshold potential, then the neuron is fired.

The molecules that are excitatory include acetylcholine and glutamate, while inhibitory ones include glycine and gamma-aminobutyric acid (GABA). In a simple sense, far too much excitation can lead to seizures while inhibition can lead to sleep, or even more extreme, coma. During sleep, wake-promoting neurons are suppressed by GABA and galanin among others, so they do not reach the threshold voltage to spike in activity. Yet, they do maintain some level of activity even when suppressed by these neurotransmitters, which is in the form of neuronal "noise," or random fluctuations in the cell potential. In wake-promoting neurons, this originates from random opening and closing in the neurons' cell membrane, which leads to tiny fluctuations in the neurons' voltage.

While a single neuron's noise is well below the necessary excitability to provoke an action potential—at least a thousand times lower—Bartsch and his colleagues wanted to examine the effects that many connected neurons could produce and if it could give rise to a spike of activity. This group effect can be explained like a bubbling sea of neurons. The neurons are all bubbling away, but suddenly their effects add up like an eruption out of the sea, resulting in the firing of some of the neurons. The team created a model to show a range of different noise values across a group of neurons.[12]

Using statistical techniques from physics, they then simulated how the neuronal noise of this group of neurons could change with time, finding that there may be some instances where the collective voltage crosses a certain threshold, resulting in some neurons firing their action potential and therefore triggering an arousal in the cortex, the outermost layer of the human brain. This would then simultaneously excite sleep-promoting neurons in the brainstem that suppress the effect to keep the arousal brief and, therefore, not force an actual wake-up.

Neuronal noise has some intriguing properties. One is that it is different from person to person, and another is that it depends on temperature. In a statistical definition of temperature, one would think that as the temperature rises so, too, do thermal fluctuations. But theoretical work carried out in 2000 showed that in a neuronal model, as the temperature increases, voltage fluctuations, or neuronal noise, instead decreases.[13] The researchers in Israel ran their simulation of the voltage on this ensemble of neurons for "low" and "high" temperature. At low temperatures, they found that the voltage crept above the threshold voltage several times and was maintained above this threshold for a long duration—which would result in the firing of action potentials. When they did the simulations at high temperature, however, they found that the threshold was only occasionally breached, and when it was, it was only for a short period of time—much shorter than at lower temperatures. In other words, the wake-promoting neurons were stimulated much more at cooler temperatures than they were at higher temperatures.

It is all very well having a nice mathematical model, but it needed some real-world data. The researchers obtained this by studying the sleep behavior of zebra fish larvae. The fish have a sleep cycle that corresponds to light and dark and they also cannot regulate their internal temperatures—much like a newborn baby. This means that the water temperature should, roughly, correspond to their internal temperature. The researchers studied the larvae over a forty-eight-hour period using a camera that tracked the movement of the fish, which were described as sleeping when they did not move.

The team studied the fish at different water temperatures, finding that as the temperature increased from 25 to 34 degrees Celsius (77 to 93 degrees Fahrenheit), the amount of sleep the larvae had more than doubled. However, the number of arousals per hour of sleep dropped from thirty-three at 25 degrees Celsius to just twenty at 34 degrees Celsius. When the researchers ran simulations in their model that matched these experimental characteristics, they found a good agreement, supporting the idea that neuronal noise is responsible for arousals.

Bartsch thinks that this model of neuronal noise could be applied to better understand SIDS. A major reason behind SIDS is that newborns struggle to regulate their temperature—after all, that is the reason why premature babies are placed in incubators. Studies have also revealed that a peak occurrence of SIDS happens in the early morning, which is when core-body temperature rises.[14] One might think, then, that the risk of SIDS is higher in the hotter summer months. Yet, SIDS is more common in the winter. The reason, it is generally thought, is that parents put extra blankets or clothes on their babies when they sleep, causing overheating and the risk to rise. One haunting piece of advice always stuck in my mind from antenatal classes: if a baby is too cold, it will cry; if it's too hot, it might not.

Babies that are at higher risk of SIDS may have lower neuronal noise levels, which not only would mean that they wake less often when the temperature is warmer, but also that they wake less often to change position or perhaps move a blanket that is covering their face. Despite the simplicity of the model, not everyone will agree with the team's conclusions given that SIDS is an incredibly complex problem, and it is unlikely that a single idea like this can solve the issue. But if the model is right, then it would at least explain how reducing the temperature of the room or the incubator could help spur more arousals and therefore help to negate the threat of SIDS. Bartsch says that a lot more work is needed, and he now plans to study premature babies in neonatal intensive care units to analyze the effect of temperature on arousals.

"After I published the work, I had a lot of emails from SIDS researchers who say the condition instead has to do with this or that," Bartsch said. "I'm not going to disagree, but we have provided the first model to connect SIDS with temperature and how SIDS could emerge."

A major challenge in advancing our knowledge of the infant brain is the difficulty of not only tracking brain activity in babies but also trying to understand what it all means. When our second child, Elliott, was born, I witnessed firsthand how complex, and messy, a newborn's brain activity can be. Having a science background and being an overly keen dad-to-be when my wife was pregnant with Henry, I was ready with pen and paper to start recording the contractions as they happened. We were told they would increase in both length and frequency, so that when they became around two minutes apart, it was generally time to call the hospital. This worked the first time around and we arrived at the hospital within good time. With Elliott, however, I thought

I would do it a bit more high-tech and downloaded a cell phone app to keep track of the contractions. After all, I could play about with the data following the birth and see if it could predict the time when the baby was born.*

Alas, Elliott did not comply. Contractions arrived one late afternoon, but after a few hours of taking recordings, it was obvious that the birth was not progressing—the gap between contractions would decrease nicely, only to then get longer. I was a little puzzled. Hours later and with no further movement, but contractions still ongoing, we decided to head to the midwife-led birth center. Once there, the birth was still not progressing, and after another four hours we were sent to the local hospital.† Following a thirty-minute drive across town, with movie scenes of car births playing in my head, the midwife at the hospital determined the location of the baby.

She found he was not spine to spine (i.e., maternal spine and baby spine aligned), but rather "spine to side," which was offered as a potential explanation for the slow progress. The contractions might not be properly forcing the baby's head to shorten the cervix. Whatever the reason, eventually, things began to progress—perhaps aided by a change of fetal position—and my wife entered the second stage of labor and duly took to the birthing pool, gas and air in hand, once the initial pain relief that the water provided had subsided.

After Elliott was safely delivered, he was placed in my wife's arms, and we both lovingly stared at him as he was covered in blood and mucus. But within a few minutes, we noticed some strange behavior. His eyes were occasionally moving rapidly, and he was lifting his arm in spasm-like episodes. The midwife observed him and felt concerned enough to call a doctor who decided the

---

* As a crude estimate, this could be done by plotting contraction duration against time. Initially the data would be noisy, but as time goes on, it would begin to converge to a single point—the time of birth. It's not possible to measure until this single point, but roughly extrapolating to it could be an indicator. Then again, it might not!

† In the UK, some midwife-led birth centers—that do not have emergency care on-site—have a four-hour rule, in which you can only stay for that length of time, but if the birth doesn't progress, then you are advised to be transferred to the hospital due to the potential higher risk of birth complications.

best place to take our son was the NICU where he could be observed. We put Elliott in a diaper and placed him in a mobile cot, and I wheeled him into the unit while my wife recovered from the birth.

It was an incredibly anxious time with a tired mind racing about what it might be and whether it could be something life-changing or even potentially life-threatening. After I went back home for some sleep, I returned to see Elliott all wired up inside the incubator as my wife sat next to him.

Five other babies were in the NICU. All of them were premature and in their own incubator. But only Elliott had numerous patches that covered his head, each of which led to bunches of multicolored wires hooked up to an electroencephalogram machine, or EEG for short. The monitor had about twenty rows of continuous tiny squiggly lines that measured neural activity in real time via the firing of millions of neurons in Elliott's brain. The doctor did not offer an initial diagnosis, but I suspect she was concerned Elliott's movements could be due to epilepsy, the world's most common chronic neurological disorder that affects about one in a hundred people.

An epileptic seizure occurs when the normal functioning of the brain is interrupted by neurons locked into a single rhythm. In this case, a group of hyperexcitable neurons at a specific location start firing in unison. This recruits other neurons to synchronize with them, which in turn engages other neurons, setting off a synchronization avalanche. The EEG is somewhat like a seismometer that measures earthquakes with the background squiggles of "noise" being normal brain activity while the earthquakes, or big spikes, could be the result of seizures.

My wife and I sat beside Elliott in the NICU watching those squiggly lines go up and down, slightly mesmerizing as they did so. Thankfully, after a few hours on the machine, the doctors were satisfied that Elliott was not having seizures, so off came the sensors. He still needed another day in neonatal care, but as time went on, his spasm-like movements in both eyes and arm faded away. We never knew what the issue was. Maybe he was tired from the delivery, as we all were.

What I learned when Elliott was hooked up to an EEG is how complex the brain's output is. I certainly could not pick out any specific "signals" on the EEG, with the lines sharply rising and falling with no pattern at all—like tracking a share price on the stock exchange. This noisy-like behavior applies when adults have EEGs, too, and so to pick out any signals, neuroscientists must first process the data. To do so, they first filter the data to remove artifacts (such as spurious signals) and then perform what is known as a "fast Fourier transform." This mathematical technique converts the EEG signal into component waves of different frequencies and amplitudes, allowing researchers to examine the different components in detail.

Neural activity for an adult is a bit like an audience clapping. Some of the time the claps are out of sync, and it results in a cacophony of noise. But then there are instances when the clapping is in sync and it generates waves of clapping, like a Mexican wave through a stadium. The brain can produce clapping waves at many different rates that can be grouped together in their own frequency bands. The most prominent rhythm in the brain comes from alpha waves, and we still do not know how they are generated. Alpha waves are synonymous with being in a relaxed state, having a frequency between 8–12 hertz. Beta waves, meanwhile, at 12–30 hertz, are produced during REM sleep and also when we are engaged in a task. Delta waves, which are associated with deep sleep, have a frequency between 1–4 hertz.

Once the component waves have been extracted from the EEG, it is then possible to determine which frequency range dominates by examining the power of those waves—the square of their amplitude—against frequency and locating any "peaks" in the data. If the power of the EEG frequency signal has a peak around 1–4 hertz, for example, then it is likely that the subject is in deep sleep.

For newborns, however, things are a bit different. One problem in studying the EEG of infants, whether they are asleep or awake, is that there is no regularity or rhythm to it; there are only short bursts of periodic activity. A lot of sleep activity in a newborn within the first four weeks of life is burst-like and transient, such as "delta brushes" that are characterized by slow delta-like waves with superimposed fast, beta-range behavior.[15] When an adult is awake in a relaxed state, alpha waves are dominant. In newborn babies, however, it is thought that alpha waves only emerge at around three months with a frequency of about 3–4 hertz, increasing to about 6 hertz at one year old. Yet, when they are first produced is still an open question. The same goes for so-called "mu" waves that are associated with voluntary movement, which are produced by the motor cortex.

This means that a newborn EEG has little periodicity lurking beneath, but that hasn't stopped scientists from beginning to examine whether there might be any signals in that noise, just as there can be in other phenomena, such as stock-market activity and heartbeat rhythms.

In the 1920s, the physicist J.B. Johnson was working at the famous Bell Telephone Laboratories in New Jersey—a lab that has a rich history in physics and that is responsible for nine physics Nobel Prizes. Johnson was studying vacuum tubes, which were first invented in the early 1900s. They were devices made from glass that contained a vacuum, or absence of gas, so that when electrodes were placed at either end, the tubes could be used to control an electric current flow inside. Vacuum tubes would later become key components in electronic circuits leading to the development of radio and TV. As Johnson worked on them, however, he discovered that there was unavoidable noise in these tubes generated by the random thermal motion of electrons.[16] It was thought the source was "white noise," in which the power or intensity following the Fourier transform is the same at different frequencies (see figure). Yet, the noise in Johnson's experiment was not white at low frequencies. Instead, the power of the signal rose at lower frequency.

The German physicist Walter Schottky investigated this low-frequency effect further, finding that the power of the frequency spectrum at low frequencies was high before gradually dropping off toward higher frequencies.[17]

It later became known as pink noise, or 1/f noise, with the 1/f referring to the inverse relationship between amplitude of the waveform and frequency. Examples where 1/f noise can be seen include stock prices, cardiac dynamics, tide heights, and even Bach's *Brandenburg Concerto No. 1.*[18] Interest in the noise in EEG signals, which also shows 1/f dynamics, was somewhat of an intellectual curiosity or, worse, ignored to focus on the "purer" brain oscillations. But some researchers—notably Walter Freeman from the University of California, Berkeley in the 1980s and 1990s—were convinced that looking at the 1/f signal could provide new clues to the brain's inner workings, for both adults and infants.

Since then, neuroscientists have developed tools to pluck out particular 1/f patterns—known as aperiodic signals—and apply them to EEG signals generated by adults and more recently by newborns and infants. In 2021, cognitive scientists Natalie Schaworonkow and Bradley Voytek at the University of California, San Diego, studied historical EEG measurements that were taken of twenty-two infants during the first seven months of life. The babies were recorded as they reached for an object, with the researchers at the time interested in the onset of motor development.[19] Schaworonkow and Voytek found a peak in the power spectrum at 7 hertz—although not present at the age of one month—that gradually grew to be much more prominent at seven

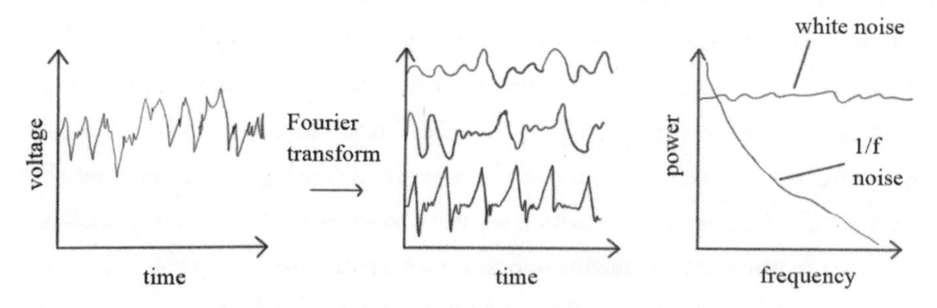

Fig 2 | *Analysis of an aperiodic signal in an EEG signal (left) results in splitting the signal up into component waves (middle), and analyzing the power of the waves with frequency.*

months, showing that the motor rhythm begins to gradually emerge in the first six months of life, similar to the onset of alpha waves.

When Schaworonkow and Voytek applied their tools to the infant power spectrum, they saw short bursts of periodic oscillations, the frequency of which gradually increased with age. But they also found large changes in the aperiodicity—related to how steep the 1/f slope is—of the EEG brain activity over the first six months. They discovered that the amount of aperiodicity in the brain decreases with age, like a "flattening out" of the aperiodic activity, as the baby gets older.[20] Interestingly, the brains of older adults tend to have differing levels of aperiodic activity than, say, younger adults, but the reason remains a mystery.[21] "It's not clear what this change in aperiodic activity means for infants," Schaworonkow said. "To find out, you would need to link the activity to behavior, which is a task for future research." Schaworonkow added that it is important to get a proper handle on the amount of aperiodic activity in the infant brain before it is possible to accurately define the onset of oscillatory behavior in infants.

Part of the difficulty linking activity with behavior is due to limited brain-imaging techniques that can be used on newborns and infants, which means that relatively little is known about the maturation of brain function at such a young age. Techniques tend to have either good temporal (time) resolution, like the EEG, or good spatial resolution, like functional MRI. Developed in the 1990s, fMRI measures small changes in blood flow that occur with brain activity. Considered one of the best brain-imaging techniques when it comes to spatial resolution, fMRI has proved invaluable for understanding brain function. However, fMRI is costly, requires large machinery, and is particularly unsuitable for young children and especially newborns given that they need to stay completely still, which, as any new parent knows, is impossible.

New techniques being developed by interdisciplinary teams could soon provide further insights into the inner working of the infant brain, which can undergo significant functional and structural changes during the first decades of life. In 2017, researchers in France created a device that combined EEG with ultrafast ultrasound (capable of taking ten thousand ultrasound frames per second, compared to fifty in conventional scanners). The sensitivity of the instrument allows the mapping of subtle changes in blood flow inside brain vessels and how that correlates with electronic signals of neural activity. They used this portable, noninvasive device to monitor seizures in newborn brains with higher resolution than what other technologies could achieve. They were also able to use the device to distinguish between "quiet" and "active" sleep in two infants.[22]

As well as monitoring the electrical signal from neurons, detecting the tiny magnetic fields that are generated by brain activity can be equally powerful. This is called magnetoencephalography (MEG), which involves measuring the small magnetic fields generated at the scalp by neural current flow, allowing direct imaging of brain activity with high precision in both space and time. MEG can probe neural activity much deeper than EEG. The issue with traditional MEG systems is that they use an array of cryogenically cooled sensors in a one-size-fits-all helmet, meaning that such systems are bulky and, again, require patients to keep very still.

In 2019, physicists at the University of Nottingham, collaborating with neuroscientists at University College London, built a MEG based on incredibly sensitive magnetometers called optically pumped magnetometers (OPMs). The advantage of OPMs is that they can be placed closer to the head than conventional sensors, increasing their sensitivity. The researchers created a lightweight, wearable MEG scanner that looks like a bike helmet and is adaptable to anyone, allowing patients to be scanned as they freely move around while maintaining data quality.[23] The power of this method is that the children can act naturally throughout the imaging process rather than being placed in a huge claustrophobic scanner.

As well as enabling studies of neurodevelopment in childhood, this system could allow investigation of neurological and mental health conditions in children, such as epilepsy and autism. The team used the device to measure brain activity in children who were performing everyday activities, including a two-year-old and a five-year-old watching TV while their hands were being stroked by their mother. The possibilities offered by these techniques in the next few years will be endless. Improvements in sensitivity and spatial resolution of such methods could allow scientists to monitor how a child's brain function changes as never before. And that is not only as they sleep but also as they tackle two major milestones in life—walking and talking, which we will examine in the final two chapters.

13

ON THE MOVE

When astronauts return home, readjusting to Earth's gravity can be difficult. The longer the time, the harder it can be. Between 2015 and 2016, US astronaut Scott Kelly and Russian cosmonaut Mikhail Kornienko both spent 340 days aboard the International Space Station, where they went about their business while floating around in the microgravity environment. After the mission was over and they touched down in the steppes of Kazakhstan on March 1, 2016, the astronauts began a grueling period of readjustment to Earth's gravity. This included relearning simple tasks, such as walking in a straight line, that most of us take for granted.

Videos from Kelly's first medical and physical examination after his year-long trip show him slowly getting up from a lying-down position and then attempting to walk forward, stumbling as he did so, like his legs were made of jelly. Six hours later, he was tested again and his steps were quicker but still uncertain. Almost a day later, he looks more stable, but still wobbly. Putting one foot in front of the other for an Earth-returning astronaut seems like a gargantuan task. It's like watching the first steps taken by a one-year-old.

Making a rather dubious analogy, this readjustment to a new, albeit familiar, environment for an astronaut is somewhat like the change a baby senses once born. Prebirth, babies spend all their "lives" in the amniotic fluid, but once they enter the world, they soon feel the full effect of gravity. Newborns do not have enough strength to move on their own, being held back in particular by their head, which can be one-third of the newborn's total body weight. Gravity essentially pins them to the ground. Indeed, it would be quite a sight to see a newborn attempting to stand up straight after birth, followed a few hours later by confidently walking down the corridors of the maternity ward. Before they can take those first unaided steps, which usually begins about a year after birth, babies generally pass several "milestones," such as rolling, crawling, and standing—as well as coasting, which involves clinging onto something such as a rail and traversing it.

The first full-body independent locomotive movement, which happens around four months, is a body roll, often from front to back. It takes another five months before babies can crawl with any proficiency and really start to move around the place, at least giving tired parents time to prepare for this sudden onset of mobility. With our firstborn, Henry, we had just about got to grips with an immobile baby lying on his playmat and staring at a toy, when, at just six months, he started "belly crawling." He would lie flat on the floor on his belly and then push both arms against the floor to move—usually backward. Despite it being an arduous movement, he was still able to get across a room in a few minutes. That was soon followed by crawling—forward this time. A few weeks later, he was attempting to climb up things, particularly the stairs.

Some infants, but not all, belly crawl (for example, the army crawl) before moving on to actual crawling, in which the torso is lifted completely off the ground. No evidence has been found that there are discrete stages toward crawling. Infants employ a variety of techniques to move around on all fours.[1] Intriguingly, babies that belly crawl first tend to crawl and walk slightly earlier than those that skipped the belly stage altogether, while smaller and slimmer babies tend to crawl earlier than their chubbier counterparts. When infants finally get proficient at crawling, they usually adopt a lateral sequence of limbs (i.e., left-hind, left-front; right-hind, right-front) presumably "choosing" this because it is the most stable pattern of crawling coordination. While this specific series of moves is what nonprimate quadrupeds also do, it is different from primates that perform a diagonal sequence (left-hind, right-front; right-hind, left-front). Despite this disparity between our closest relatives, work in 2015 of seven ten-month-old babies found that many movements during crawling, such as stretched-out arms, were like that found in nonhuman primates.[2]

Crawling eventually makes way for walking (although some babies may skip crawling altogether). Before babies take their first independent steps, many parents help them along by holding both their hands and letting them put one foot in front of the other. Research has shown that babies helped with "stepping exercises" before they could walk were quicker to walk independently than those that were not given this training.[3] Although why you would want your child to walk before it is absolutely necessary is beyond me. We never did stepping exercises with Henry, but it seems like he didn't really need the help. By nine months, he had already started taking his first step. I was not there to see it happen, but my wife thankfully filmed it, and I later watched in disbelief. He was on all fours, in a high crawl position, and then rocked back to get in a squat. He then stood up and took four stumbling steps before falling forward and putting his hands on the floor. It was like watching a reenactment of the first primate becoming bipedal.*

The struggles newborns have getting around do not necessarily translate to others in the animal kingdom. If you have ever watched wildlife documentaries or programs about life on a farm, then you would have likely seen an animal give birth—be it a cow, bison, or horse. Within thirty minutes after entering the world, a calf or foal can be seen trying to stand, while it is not unknown for a foal to be running an hour later. Likewise, a wildebeest calf can move within minutes of birth and only after a few days is able to walk alongside the pack. This has led some to think that humans are fundamentally different from quadruped mammals in how they develop their walking

---

* The latest research suggests that this happened some eleven million years ago; see: www.nature.com/articles/d41586-019-03418-2.

ability—a so-called "locomotor divergence." In other words, it is thought that humans have insufficient neurological development as well as strength, to control the necessary muscle groups to get up and walk. It begs the question that if human babies were born one year and nine months following conception, would they be able to stand and possibly walk after birth? The answer is, perhaps, but it would require such significant changes to the female pelvis as well as the ability to carry such a heavy fetus for that length of time while remaining on two legs that, for human babies, walking shortly after birth is simply not an option.

What is surprising, however, is that if you do these same stepping exercises with a newborn just a few days old, the baby will also take "steps." If you support about 70 percent of an infant's body weight by holding them on the torso and then gently place their feet flat on a solid surface, they will attempt to walk by placing one foot in front of the other alternatively, just as you would. This is thought to be another newborn reflex—the stepping reflex—that gives the impression that a newborn is already ready to start walking after birth. This rather mysterious reflex disappears after around two to three weeks. But this is not quite the end of the story. While it is assumed that it takes months for an infant to move adeptly, the latest research shows that newborns can still do remarkable things. Perhaps we have more ability—and with it are closer to others in the animal kingdom—just days after birth than we might think.

For most of us, walking is something we don't have to think about—it just happens. While you might suppose that the brain plays a large part in controlling the motion of walking, instead it is carefully choreographed by the central nervous system and motor neurons in the spinal cord. The spinal cord, which runs down the back and lies within the vertebral canal, is about the diameter of a finger and the length of a thighbone. The central region of the spinal cord, called "grey matter," contains the synapses and cell bodies of the neurons while the surrounding "white matter" consists of the axons that then transmit the electrical impulses up and down the spinal cord (in a similar way to the white and grey matter that exists in the brain). The spinal cord is segmented, with each section featuring outgoing motor neurons to the muscles as well as incoming axons of sensory neurons—those that are activated by sensory inputs like on the fingertip.

This clumping together of neurons in the spinal cord into functionally related groups simplifies the activation of groups of muscles in a particular order to produce movements—say, the flexing of a bicep on the front of your upper arm or the triceps that extend your forearm. (Motor neurons also connect to organs and glands.) The clever aspect about neurons in the spinal cord is that they connect sensory neurons to motor neurons to create their own neural circuits, called central pattern generators (CPGs).

The activity by the CPGs is a bit like a conductor conducting an orchestra, making them play at the right time to create a melodic tune (i.e., smooth walking). These circuits can also spontaneously produce a repetitive output with no sensory input, which relieves the brain from coordinating certain movements, such as walking, chewing, or breathing, so it does not have to constantly specify how the motor neurons need to act. Of course, the brain still needs to initiate the action, but the details are left to the CPGs. The ability of these neural circuits is perhaps best demonstrated by decapitating a chicken and seeing it run off, even flapping its wings as it does so; or, for a more mundane example, why it is so easy to concentrate on a podcast when out for a walk or a run. Your brain can focus on the content of what you are hearing without being diverted by the actual process of movement. There is still "higher level" input, however, from the brain when walking or running to allow the body to undertake certain movements, such as rapidly changing direction or dodging moving obstacles.

In 2011, researchers in Italy and the United States investigated how walking ability via the CPGs develops from birth to adulthood. They enrolled forty-six newborns between two and three days old and used electromyography to analyze how the nerves stimulate the leg muscles by measuring the electrical activity while the babies moved their legs during the stepping reflex. The team used mathematical techniques to elicit patterns in the electrical activity as the infant moved using twenty or so different skeletal muscles. They found that as the newborns took a step, spinal-cord neurons were activated in two specific patterns—one that commanded the legs to bend and extend while the other made the legs alternate in steps.[4]

When the team then studied ten walking toddlers, they saw something different. In this case, there were four distinct patterns. However, two of them were still the original stepping patterns seen in newborns. The two additional patterns controlled more nuanced aspects of walking, such as when the foot touched down as well as when the foot pushed off the floor. These aspects allow the toddlers to control the muscular force they use to walk and the speed of the movement. With more confident toddler walkers, this four-phase pattern became stronger until it resulted in a "mature" pattern of motoneuron (or motor neuron) activation that is seen in adults, in which each phase is effortlessly timed at a specific part of the movement.

What is fascinating is that the stepping reflex seen in newborns is not locomotion itself but, from a neurological point of view, it is a building block toward that first independent step.* In other words, rather than the neural patterns of the newborns' stepping being discarded after a few weeks when the reflex disappears, it is rather retained and tuned with new neural circuits added when later learning to walk.

As every parent knows, newborns show another type of leg movement at birth—kicking spontaneously in the air. As we learned, babies start kicking in the uterus and continue to do so after they are born until they are around four months old or so. The most frantic kicking usually happens just as you are trying to put a new diaper on the baby, making it nearly impossible to do so at times. It had long been thought that newborn stepping and kicking were identical movements generated by the same neuronal mechanism, but some of the latest work is challenging that assumption, finding that kicking and

* The four-phase pattern is also seen in many quadrupedal mammals, such as rats, cats, and monkeys, as well as in birds such as guinea fowl that are bipedal, showing that similar neural patterns are at play across different vertebrates.

stepping are produced from different neural processes.[5] Yet, understanding how babies generate these kicks is difficult, as a detailed understanding of individual nerve cells has not been possible without undergoing surgery.

Previously, measuring the electrical signal from the leg involved sticking small needle electrodes into the skin at various points. In 2020, scientists from Italy, Germany, and the United Kingdom created the world's first noninvasive way to map how baby movements are generated on a single motoneuron level. The researchers developed a high-density electrode cuff that is worn over the whole of the lower leg, and using a mathematical deconvolution technique, they deconstructed the many signals from all the motor neuron firings so they could track the behavior of thirty individual motoneurons. According to Francesca Sylos-Labini at the Santa Lucia Foundation in Rome, this technique is like setting up a bunch of microphones at a cocktail party to pick up what a single person is saying among a group of people who are all talking at the same time. By applying this mathematical trick, scientists could extract individual motoneuron signals among many others that were firing at the same time.

When the team monitored the activity of the motoneurons in such a manner, they found that, unlike in fast-leg movements in adults, the babies' kicks were generated by neurons in the spinal cord firing at the same time—what the researchers call "extreme synchronization."[6] This effect had already been seen in rats, but scientists were unsure if it also happened in humans. Indeed, the work goes some way to explain why babies' kicks can be hard and fast even though their muscles are relatively weak. The team adds that all this evidence points to the importance of CPGs in how newborn babies kick and step. "This gives further evidence that the brain, or cortical, control of the newborn's leg movements is low, unlike in adults," said Sylos-Labini, who has a PhD in biophysics.

The researchers are now applying their findings to not only improve the monitoring cuff but also to test whether it might be able to spot early signs of motor disorders. One is cerebral palsy, which is often not diagnosed until the infant fails to move as a normal infant would, which sometimes can only be determined when they are many months, or even a year, old. Infants born extremely preterm have a 10 percent risk of developing the condition. The researchers plan to investigate the locomotor neural patterns generated by preterm infants to see if they can pick out any biomarkers. If anomalies in locomotor ability could be found early, then rehabilitation could take place sooner to possibly lower the impact of such conditions. "The brain is incredibly plastic in the first few years of life," said Sylos-Labini. "Having an early diagnosis could make a huge difference in therapy."

What all this research shows is that, through kicking and stepping, the basics of neuronal control for bipedal locomotion are already present at birth—they just get tuned and improved upon until the baby takes those first steps some nine months later or more. But what about quadrupedalism, such as crawling? Is that something an infant learns after some months of rolling around on the floor, being solely a milestone on the way to walking, or is it perhaps something also innate to who we are?

One of the most iconic album covers of all time is Nirvana's 1991 *Nevermind* album. The artwork shows a baby underwater looking at a one-dollar bill attached to a hook with its mouth open and arms spread wide. As legend goes, the idea was conceived by Kurt Cobain after he became fascinated by a television program on water births. My second son, Elliott, was a water-birth baby, and it never dawned on me to tie a one-dollar bill to a fishing line to help coax him out. Anyway, it is well documented that when put underwater momentarily, infants will attempt to "swim" by kicking their legs and moving their arms (although they do not have enough strength to actually swim). Someone who has been fascinated by how babies move, whether "swimming" in water or attempting to crawl up to the breast to feed, is Marianne Barbu-Roth at the University of Paris Integrative Neuroscience & Cognition Center. Barbu-Roth studied nuclear physics before earning her PhD in embryology, investigating the movement of cells in the embryo particularly during the formation of the neural tube, which eventually morphs into the brain and spine.

After continuing in biology for a decade, Barbu-Roth concluded that dabbling in genes was not for her and decided to change direction. Part of that switch came after seeing videos of newborns attempting to move and becoming intrigued by their abilities. Discovering a link between this and her previous work on cells, she wanted to study this aspect of movement further. "After all, everything needs to move," she said. "In a cell, for example, I found it fascinating that the movement must be done at the right time, in the appropriate 'time window,' and wondered if this applied to newborns. Do they need to do certain movements in certain time intervals to progress?"

In the late 1990s, Barbu-Roth moved to the University of California, Berkeley to work with developmental psychologist Joseph Campos, a pioneer in the study of infants in "optical flow," a form of visual streaming that occurs when moving continuously in one direction. (Imagine sitting in a flight simulator. While you yourself are not moving, the screen gives you the impression that you are.) How the eye interprets optical flows is critical to how adults control movement, and in the 1990s, Campos and colleagues showed that such flows are processed differently when infants learn to crawl. They found that the ability to use optical flows in peripheral vision to control actions such as crawling is not innate. Pre-crawling infants do not have any fear of heights (probably best not to test it). Yet, it was an open question whether newborns showed any type of response to optical flows.

At Berkeley, Barbu-Roth met David Anderson, who has a background in kinesiology and is now at San Francisco State University. The pair joined forces to study newborn responses to optical flows. Some of their early work focused on infant stepping, in which they discovered that newborns tend to step more in the presence of an optical flow than when a static visual was presented.[7] The duo suggest that, given the movement was linked with vision, it indicates that there must be a higher-level, or "supraspinal," control when stepping. In other words, stepping is not a reflex at all.

More recently, the pair moved onto crawling, and when Barbu-Roth returned to Paris in the early 2000s, her team installed a tabletop LCD screen in the lab that showed random scattering of black dots on a white background.

The team enrolled twenty-six newborn infants and placed light-reflective sensors on their joints to record the quantity and quality of their movements. When the infants were placed on the screen, they would move their hands and legs just as they would on any surface. When the researchers made the dots flow in one direction, this gave the three-day-old newborns the illusion that they were moving. The babies would then move their arms and legs significantly more than they did in the static display.[8] The same effect happened when the newborns were held above it in the air. They would try to move in a crawling-like way, like swimming in air.

Barbu-Roth and colleagues then wanted to examine if a newborn had the ability to move themselves around by crawling. This was trickier to test, of course, as newborns do not have the strength to move independently. After some thinking, the team designed what is effectively a mini skateboard that they could place the baby onto to fully support them—especially the heavy head—but still allow the baby to use its arms and legs to move freely. The team tested sixty-two day-old newborns and found that most babies were able to crawl using the device—called the Crawliskate*—and could easily traverse across a table, propelling themselves along using only their arms and legs. When the researchers analyzed the movement of the leg via markers with a camera, they found that the patterns were similar to those documented during quadrupedal locomotion in adults and animals.[9]

In 2020, Barbu-Roth went back to answer a question that first instigated her interest in newborns in the first place: the impact that milk, either smell or taste, has on the infant's ability to move. Barbu-Roth and colleagues infused the head pad of the Crawliskate with either the mother's breast odor or the odor of water, which served as the control.[10] Analysis of the number and types of limb movements, as well as how far the newborns traveled across the surface, showed that infants are significantly more efficient crawlers when they smell maternal breast odor. Despite performing fewer individual movements, they were able to move greater distances, which again shows that higher-brain

---

* The patent for the device was granted in 2016: patentscope.wipo.int/search/en/detail.jsf?docId=WO2016009022&tab=PCTBIBLIO.

processing is occurring given that the infants were moving in response to smell. The research also shows that maternal odor could be used to help premature babies foster and build neurological connections, given they are at risk of developing certain neurological conditions. Barbu-Roth and colleagues are now planning to investigate what other senses may play a role in movement. With the help of the Crawliskate, they are studying the effect of the maternal voice on a newborn to see if it has the same impact as breast odor in terms of increased movement (which they think will turn out to be the case).

Anderson says that perceptual research on newborns indicates that not only do multiple sources of information play a fundamental role in newborn movement control but also that this control must already be localized in higher brain centers. In addition, Barbu-Roth and Anderson have proposed that bipedalism not only emerges from quadrupedalism but that what we view as bipedalism is in fact quadrupedalism because all four limbs are engaged in an activity such as walking. In other words, in the nervous system we are quadrupeds when first born, but we never lose that quadrupedal organization even when we develop bipedalism. The researchers think that crawling should not be seen as a stage toward walking but something more integral to how humans move in early life and beyond—a view that is certainly controversial and not widely agreed upon, but according to Anderson, one that is gaining increasing traction in the community.

"Newborns have a major physics problem when born in trying to move against gravity," Barbu-Roth said. "But in the nervous system, they certainly have the capability to move." Indeed, if a human were ever to be born in a microgravity environment—sometime in the distant future, perhaps when humans start to colonize other planets—a newborn might have a better chance of getting around than it does on Earth. Just make sure to have some breast-odor-infused rags to help guide them around the place.

While crawling could be more integral than we think, it will only get a baby around so far, even when being propelled across the room on a Crawliskate. To really get about, walking (or running) is eventually the best option. As we discovered, when adults walk, it is a repetitive and precise movement. If you take a stride with your right foot, you may notice that the body's center of mass (located just under the navel) dips slightly as measured from the floor when you do the splits. When moving your left foot forward, the center of mass rises again as the foot passes by the right foot for an instant, before dipping again when the left foot is now forward, with the cycle repeating (see figure).

This is known as the inverted pendulum mechanism, or upside-down pendulum, and has been demonstrated in many species, such as birds, elephants, crabs, and cockroaches, despite the wide variety of body sizes and skeleton types.[11] The reason why it is employed so extensively in nature is because it is an incredibly efficient form of locomotion, the efficiency of which can be explained in terms of mechanical energy—the conservation of kinetic potential as the person moves with the potential energy change as the center of mass moves up and down. The amount of energy recovered depends on the speed of walking, but for the optimal case of about 4.5 kilometers per hour (about 2.8 miles per hour), it can be as high as 65 percent in adults.[12]

Some of the early work in the inverted pendulum mechanics of walking children was conducted in the early 1980s by researchers in Milan, Italy.[13] They studied forty-two children aged between two and twelve who were made

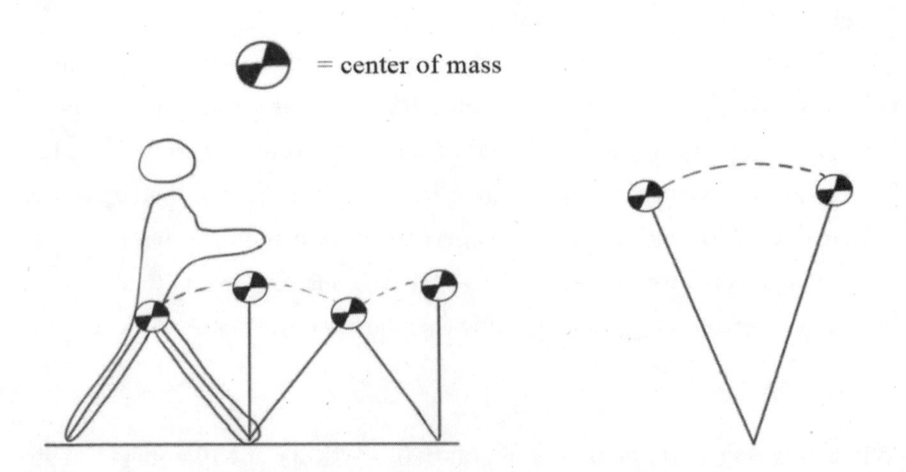

Fig 1 | *When we walk, our center of mass undergoes an inverted pendulum.*

to walk at different speeds over a strain-gauge platform, a device that measures the applied force by the feet when moving. The study found that the optimal speed—resulting in maximum energy recovery—for a two-year-old is 2.8 kilometers per hour (about 1.7 miles per hour), and increases progressively with age until reaching the adult value when the child is twelve. When a child chooses the speed at which to walk, the energy recovery roughly matches the ideal optimal speed. The researchers showed that when a child travels faster than this optimal speed, the energy recovery decreases, and the younger the child, the greater extent the energy loss is. What this means is that the body must do more work to compensate for the reduction in energy efficiency. So, when you are badgering your young children to walk home faster from school to catch the start of the football game, it really is more tiring for them.

Initial movements for infants who are taking their first steps usually involve short, quick steps and rocking the body back and forth for balance. The baby's toes point outward with a wide support stance, and the arms take up a "high guard" position. Even when supported by an adult, toddlers still show the same walking pattern, highlighting that this posture does not emerge from being unbalanced. Despite an infant's first steps appearing like someone who has just been thrown out of a bar, it is effective at stopping falls. In 2004, scientists applied the mechanical analysis techniques of walking to the first unsupported toddler steps.[14] Twenty-six children between eleven months and thirteen years were filmed, with the eight youngest children recorded daily so that the researchers could capture those first steps.

Placed on a force plate, the infants wore several infrared reflective markers that were tracked with a camera. The researchers found, as in previous studies, that adult walking resulted in an optimal energy recovery of about 65 percent, but for toddlers it was much lower, at 35 percent. Babies taking their first steps lacked the pendulum movement completely, instead sacrificing energy efficiency (which was estimated to be as low as 25 percent) to maintain upright stability. So, as perhaps expected, infants' first attempts at walking have little in common with how older children, those from four and above, or adults walk. This led the team to conclude that the pendulum motion is not an innate consequence of the motor function and environment, but something that is learned through the walking experience, taking about two months to bed in.[15] Indeed, this initial lack of a pendulum behavior has also been found in newborn chicks during the first two weeks of life.[16]

The simplicity of the pendulum motion hides the fact that walking takes a large amount of neural control—effectively controlling a trajectory around a virtual point, the center of mass. The high variability in certain movements when learning to walk could reflect the central nervous system's exploration of a wide range of "solutions" before somehow settling down on a common optimal one. Indeed, studies have also found that there was no difference in walking behavior for infants that were supported, which suggests that the idiosyncratic features in walking toddlers do not result from underdeveloped balance control but rather an innate template of stepping.[17]

Of course, infants do not learn to walk on a treadmill or in a perfect laboratory setting. Instead, they do so at home where there are objects lying around the floor, uneven surfaces, and things to hold on to. Scientists have looked at what happens in a laboratory setting when an obstacle is put in the way of the developing walker. When walking down an inclined plane—a favorite for any physics high-school demonstration—adults would walk up and down at roughly constant speed. Infants, on the other hand, lack this control and will increase their speed when going downhill and decrease it when going up.[18] If a small object was placed in their way, about half of the infants would just stand on it—hopefully, it was not a LEGO piece—with about 25 percent stepping over it while the remainder either stopped or just stumbled into it. When it came to stairs, toddlers liked to put their foot on the edge of the step rather than the flat surface, something that was more prominent when coming downstairs. (The adults tested under the same conditions never stepped on an edge.) The conclusion was that both the object and the stairs represent "haptic" probing (haptic relating to the sense of touch) without the need to move the body forward.

Yet, that was still in the lab. So, in 2019, researchers in the Czech Republic and the United Kingdom decided to see if the home environment could shed any light on how infants learn to walk.[19] The researchers scoured YouTube for footage that parents had uploaded of babies' first steps. After coming across ten videos of novice walkers between seven and twelve months and ten improver walkers aged between eight and thirteen months, the team analyzed their movements, such as how the infants walked and how many times they stopped or fell. They discovered that in a home environment, novice walkers still fell more than their developed counterparts as well as spent more time with their feet touching the ground, but there was not a reduction in stops, defined as no movement for more than four seconds.

The improver walkers also often started walking on their own accord and carried something in their hand—an aspect that has not been replicated in lab experiments. The team also found, as have previous studies, that, with improved walking confidence, the number of posture changes reduces as the baby fine-tunes its walking pattern. At the onset of walking, infants are just learning to maintain an upright stance to balance against the force of gravity, while in the second stage, they fine-tune their walking patterns. And as every parent knows, after taking those first steps, infants learn to walk incredibly quickly, and before you know it, walking turns into running, which can bring its own set of challenges for both infants and parents.

Travel broadens the mind, whether traveling into space or to another country, or a baby simply moving across the room in the house. The latest research shows that newborns have more capability to move than we ever thought possible. As infants gradually obtain the ability to explore their surroundings, it is the transition between quadrupedalism and bipedalism, which is a complete game changer. Suddenly their world opens, allowing them to explore it as never before. And it is not just locomotion that is changing. Taking those first steps to explore the world is such a big milestone for both parent and infant—a sign that the baby is not really a baby anymore. Indeed, parents even change how they communicate to their child once their little one starts walking.[20] The start of independent walking not only lets them explore but is also timed with the rapid development of language, the greatest cognitive ability that humans possess.

FOURTH INTERLUDE

PHYSICS OF THE PLAYGROUND AND TOYS

One of the best aspects of having children—yes, there really are some—is going back to your own childhood and heading to the local park on a regular basis. Given that children love going to the park, it is an activity that will not take much cajoling. But why do they enjoy it so much? All parks worth their salt have a slide, swing, merry-go-round, climbing frame, and possibly even a trampoline. What the equipment allows is for children to experience a range of forces in a hopefully safe environment, be it circular motion, angular momentum, energy transformation, or simple-harmonic motion.

Perhaps the first apparatus a child will experience is the swing, which even babies that are strong enough to hold their necks up can enjoy. The swing is a simple pendulum, like in a clock that swings from side to side in a periodic way, the amplitude of which decreases thanks to air resistance. The swing converts potential energy (at the top) to kinetic energy (at the bottom). The speed of the swing is therefore greatest at the bottom and zero at the top. While velocity is greatest at the bottom, it is rather the acceleration—described as the rate of change of velocity—that is greatest at the top of the swing.[1] What this means is the swing allows the rider to feel both heavy (at the bottom of the swing) and light (at the top) over and over again—ad infinitum.

When children get a bit older, they will learn to swing by themselves and finally give their parents a break from pushing them for minutes that seem like hours. What the child has learned, perhaps with some teaching from their parent, is to "pump the swing," but what is the best way to do this? A physicist would do it in a simple way, which takes just one simple movement, but in this case, you would need to stand and not sit on the swing. Stand on the swing and then get it going a little by either someone giving you a push or rocking yourself backward and forward. Then get into a crouch position, but when the swing reaches the bottom of the swing frame, stand up quickly, and after a few iterations of crouching and standing only when it is at the bottom of the frame, you will find that the amplitude of the swing is growing through a pumping action.* By standing up quickly in such a way, you are changing the center of mass of the oscillator, and given this is timed when the centripetal force is highest (i.e., at the bottom), it produces an extra boost that increases the oscillation speed.

Of course, this is not what you see children at the park doing—although children do like to stand on swings. They have been taught another way. This involves rocking the swing. On the top of the backswing, seated riders rock backward on the seat and thrust their feet forward so that they are almost horizontal. Then, on the downswing, the rider lies horizontal (negating air resistance), pushing the swing forward to a higher point, while at the top of the forward swing, riders sit up, perhaps rocking the swing forward slightly at this point. (If I have not described this very well, I guess you know in your mind how someone rides a swing.)

In the 1990s, physicists studied the mathematics of this more complex seated pumping action.[2] They found that it can be described mathematically as a "driven harmonic oscillator." A harmonic oscillator is one that, when displaced from its equilibrium position, experiences a strong restoring force to that equilibrium. The "driven" part means that the oscillator is driven by

* This is known as a "parametric oscillator," a type of driven harmonic oscillator.

an external force (i.e., the rider). The transfer of angular momentum—a rotational form of momentum—from the rocking movement of the swinger to the swing oscillation is why the swing increases its amplitude. This can be shown in a simple experiment in which you have a wheel hanging from a fixed point, but then have a piece of string that spins the wheel (therefore adding angular momentum). If you slightly spin the wheel at the top of the swing, both backswing and forward swing, then the amplitude of the swing oscillation increases. Simulations in 2005 showed that getting the best push to pump the swing involves rocking at the swing turning points as quickly as possible—as you'll see most people do if you watch them at the park.[3] In a similar way, it is possible to pump the swing in this rocking manner while standing on it (without the need to squat).[4]

So, which one is best? Thankfully, physics has the answer. Three physicists in the United States in 1998 analyzed the two different swing techniques: seated rocking and standing and squatting. They found that seated swinging is best when getting the swing going, but once it is, then standing and squatting produces much bigger swinging amplitudes.[5] At the park you may see older children do this, without them even knowing anything about differential equations. Another common piece of equipment in every park is the slide. This is not just a case of potential energy at the top of the slide converting into kinetic energy as you slide down. As anyone knows, sometimes it can take ages to come down, not only due to air resistance, but particularly because of friction—the force between two surfaces as they slide across each other. You cannot do too much against air resistance other than lying on your back or front, which is often carried out by children once they get bored of sitting up. But if you really want some fun, then friction is what to tackle, or rather, reduce. The best way to lessen friction almost entirely is to put water on the slide and go down naked, which can create hours of fun in the garden with a small paddling pool and, hopefully, no accidents occurring or trips to the emergency room. If you are in a public park, this is unlikely to be an option, unless you want some unwanted stares of disapproval from other parents. But there is another way—wax. Get some kids to write their name on a metal slide in big letters and then, after a few have smeared it around a little, it can rapidly boost the "performance"—or reduce the time taken to descend, and hence the speed, by about 20 percent. If you want to test this out, then you might want to use a light crayon or use some clothes you do not mind getting destroyed.

My children are still obsessed by the merry-go-round, so much so that I worry they will be on it long enough to pass out. The merry-go-round is all about the conservation of angular momentum—a measure of the amount of rotation an object has. The key to the merry-go-round is that it is a heavy structure, with its mass distributed some distance away from the point of rotation that is in the center. The merry-go-round's moment of inertia—a measure of how much a body resists being rotated—is much bigger than that of the passengers', so the combined moment of inertia is little changed as the passengers move toward the center.[6] This means that the centripetal force on the riders changes as they move toward the center of the merry-go-round. As the passenger moves toward the center, the radius decreases and so does the centripetal force for a given angular velocity, while the opposite happens when the rider is closer to the outer edge. So, when putting babies on for a spin for the first time, it might be best to put them more toward the center of the merry-go-round rather than on the outer radius where they might enjoy it more, but also might fling off on a tangential path.

For older children, there is another way to experience rotation at the park: the spinner. This is a stationary pole with a small circular platform to step on near the bottom of the pole. Sometimes the spinner is tilted at an angle as it spins on its axis. The difference here is that the pole is usually lightweight with little mass distributed away from the center, resulting in a small moment of inertia. If you stand on the spinner and push your body outward to tilt backward by stretching out your arms (while still holding on to the pole with your hands), then the moment of inertia is large. But when you then bring yourself closer to the center of rotation, the moment of inertia dramatically falls. This conservation of momentum results in a large increase in angular velocity, or spinning speed. It's the same thing that figure skaters take advantage of when they spin at incredible speed. They first start with their arms outstretched (larger moment of inertia) and then quickly bring them inside their body to spin around faster.

One playground apparatus you might see in a park, but will more likely see in a lot of people's backyards, is the trampoline. This is all about the conversion of potential energy into kinetic energy, as we have seen in other examples. But there is an extra key ingredient: the springs in the trampoline that exert a force on the person, propelling them into the sky. This force is due to Hooke's law, named after the British physicist who devised the theory in the mid-1600s. The upward force is equal to the stiffness of the spring multiplied by how much the spring or trampoline material has been displaced by the jumper. So, when trampolining, the user's potential energy turns into kinetic energy as they fall, and this converts into spring energy—with the bigger the displacement, the bigger the force pushing you back up.

If you want to maximize the size of your jump, then the trick here is to increase your center of mass. This is mostly done by swinging your arms out straight above your head—which you will see professional gymnasts do. The reaction of the springs is strong, as it quickly decreases the speed of the jumpers coming into contact with the trampoline and then thrusts them back into the air. So, how large is this force? Experiments carried out in 2015 at both a children's trampoline in a park and a gymnastic trampoline showed that, for a split second, the *g*-force (which we came across in chapter seven) can be as high as 7 *g* in a children's playground—equivalent to seven times the force of gravity humans are normally exposed to when on Earth—and 9 *g* on a gymnastic trampoline.[7] The bottom line being that if your kids like to experience extreme forces—albeit for a fraction of a second—then the trampoline is what will generate it.

Which at least goes some way to explain why children pester their parents so much to have a trampoline in their backyard.

⚛ ⚛ ⚛

In a similar manner to going to the park, having a child lets you relive your childhood as you rummage around in your parents' home/attic/shed for those long-lost toys you have not played with for some twenty or thirty years.

While children these days are obsessed by tablets and other screens, you have a few years before they enter the world of *Minecraft* to show them all your old favorites.

The spinning top is perhaps one of the oldest (known) toys—one that has been enjoyed by children since the time of the ancient Egyptians. The physics behind this gravity-defying toy, balancing on a tiny point, is based on angular momentum. The angular momentum of an object only changes when you apply a torque (or, say, twist it). You can think of it as being like linear momentum, which only changes when a force is applied, like a push or a pull. So, when the top is spun, a large torque is applied that imparts a lot of angular momentum. In full flight, the spinning top easily remains upright—sometimes lasting for minutes—as the torque, due to gravity, is not strong enough to change the top's movement. However, as the top slows down due to air resistance, gravity takes over and the spinning top eventually crashes to the floor. There are many different types of spinning tops. Some are magnetic and levitate in air as they spin. Others are a transparent hemisphere (like a see-through teacup) that have small balls inside. Once in precession, the inner balls circle around the rim of the hemisphere, following the rotation.[8]

Another classic toy is the Slinky, which consists of a helical spring made from plastic or metal. The Slinky has been enjoyed for well over half a century, thanks to its amazing ability to "walk" downstairs unaided. It was created in 1943 by Richard James, an American mechanical engineer who was working on designs to keep equipment secure on ships as they rocked around on the rough seas. As the legend goes, he was playing around with coiled wires, and as he dropped it, he watched it tumble end over end on the floor. Ever the inventor, James was intrigued by this, and he began to wonder if it would make a good children's toy. After playing around with the idea in 1945, he and his wife, Betty James, founded James Industries. Two years later, he was granted US Patent 2415012 for the Slinky, described as "sleek and sinuous in movement or outline." Hundreds of millions have been bought since, and the toy took other forms, such as the Slinky Dog—made famous by the *Toy Story* films—which was also patented by James.

The Slinky is a simple toy, and while the motion behind how it works might look easy enough, the physics behind it is complex. Every object has potential, or stored, energy. When placed at the top of the stairs, the Slinky will, of course, stay perfectly still. But as you set it on its way, by pulling one end onto the next step below, this potential energy is transferred to kinetic energy. As the Slinky coils on the step below, this kinetic energy propagates like a wave through the first part of the spring to the other end. Much like when you crack a whip, all the energy travels through to the end of the whip. This "pulsing energy" then makes the Slinky spring down to the next step, and the process continues until it runs out of steps or collides into the wall.

Finally, many families have a wooden train set in which the connector pieces are joined together with magnets. I remember watching Elliott trying to get a carriage to connect to the main train whose magnets were opposite poles. To a child, it must have been incredibly confusing why the pieces magically snap together for a particular orientation but not for others. Elements like iron are magnetic, in that all the electrons of the iron atoms are "spinning" in the same direction, which magnifies the effect of a magnetic force. This force can be felt over a large distance. (I used to have "fun" trying to see how close I could get one connector piece to another before I felt the strong magnetic attraction.) There are many terrible explanations for how magnets work, so I am not going to add to them. All I will say is that, having a PhD in magnetism myself, I can't tell you much about how magnets work because we still don't fully understand them.

Maybe your little one will work it out some day.

14

TOWER OF BAB(Y)EL

When my two children were newborns, I really struggled with "parentese." This is the mostly annoying way (at least I think) of talking to your baby that involves saying words that either end in "y" (like bunny or doggy), sound like their meaning (woof or splash), or have repeated syllables (for instance choo-choo), all the while in a high-pitched voice. My mantra was to instead talk to my two little ones as I would to an adult. That might sound incredibly boring and dull, and, according to the latest research, it is also perhaps the wrong thing to do.

It is thought that parentese, being slower, variable in pitch, and more animated than general talk, acts like a "social hook" for babies, inviting them to respond. In 2019, linguists at the University of Edinburgh found that infants whose parents used more instances of baby talk learned words quicker than those babies that did not get so much cooing.[1] The researchers recorded samples of speech spoken by an adult to their child and analyzed it for "baby talk." They then measured the language skills of forty-seven infants at fifteen- and twenty-one months old, seeing a boost for infants that were talked to using more y-ending words and repeated syllables. However, not all types of baby talk saw a similar boost: There was no benefit when saying words that sound like their meaning, like woof.

Whether in parentese mode or not, as I talked to my sons when they were babies (who are two years and five months apart) and watched their responding gazes and movements, I could almost feel the firing of neurons in their brains as they took it all in. At this age, their brain is on fire. Not literally, of course, but all this listening and observing is done for a reason—to soak up information and language like a sponge. When my firstborn, Henry, was near his baby brother, I noticed how much Elliott smiled. He was transfixed, listening to everything that Henry said and watching his brother's every move, sometimes letting out a giggle, despite being occasionally poked in the eye/mouth/ears.

Researchers at McGill University in Montreal investigated this rather curious form of peer adulation by using a synthesizer to imitate vowel sounds of humans at any age.[2] The team played audio clips of such vowel sounds as would be spoken by an "adult" and an "infant," finding that babies spent 40 percent longer listening to the higher-pitched sounds generated by infants. Not only did the babies dwell for longer on the infant sounds but the sounds also prompted them to smile or giggle, just as Elliott did with his older brother.

Newborns recognize their own language well before they can perceive its features, such as vowels and syllables, and they accomplish this by becoming attuned to the sounds of what they hear. It is even suggested that the melody of babies' cries is shaped by their native language.[3] In 2009, researchers analyzed the crying patterns of thirty French and German newborns, finding that the French group produced cries with a rising melody contour whereas the German group produced falling contours. It is thought that infants as young as four months can recognize their name and two months later can understand nouns, suggesting that they can link words and objects from an early age.[4] Infants at six months also start to babble in syllables like ba-ba and da-da. But these combinations of vowels and consonants do not have any meaning, so no need to run to your partner claiming they have said their first "dada." They are likely just babbling, experimenting with the sound of language, rather than trying to get your attention and fill you with joy for the rest of the day.

It is difficult to put into words just how hard, in principle, it is for babies to pick up their language, but how remarkably good they are at doing so. This speed of language acquisition is far from being fully understood, and how babies learn the building blocks of language is hotly debated, with many competing theories. Babies discover individual words while listening to other people, but to them, language is initially just a stream of noise. Where do words start and end? Babies must crack this system as they listen, to not only break words into "phonetic categories"—the grouping into vowel- and consonant-like sounds—but also break up sentences to represent those individual words.

Some researchers have suggested that infants group sounds into native consonant- and vowel-like phonetic categories by a process of elimination akin to "statistical learning."[5] English-speaking infants, for example, can often distinguish "r" and "l"—as in "rock" and "lock"—before they are a year old. But Japanese speakers, when hearing these two words through the filter of their native language, often confuse the two, as the phonetic production of those letters can be realized by the same Japanese consonant. Other research has found that babies listen to the melodies of language, carving out the individual units of language, such as vowels or syllables, as they do so.[6] This could also be how they learn multiple languages given that languages have different rhythms, allowing babies to differentiate between them by using their musical properties.

While an infant is just coming to grips with language early on, by the end of the first year, something magical happens. Their babbling becomes more complex with a variation and strings of syllables to produce da-da-dee or ma-ma, and it is thought that infants already "know" between ten to fifty words, despite not being able to speak, and some may even understand word combinations. In 2021, researchers assessed thirty-six infants who were about eleven months old, analyzing their language-learning behavior in a series of attention tests using recorded adult speech. They found that infants who are still learning single words are simultaneously also acquiring word combinations, such as "clap your hands."[7] By the time infants turn one, around 75 percent of them will have uttered their first word.[8] With Henry, we listened avidly every day for his first word. Despite offering him much encouragement to say "mama" and "dada," his first utterance was actually "nana" (banana). It is unsurprising that food came first—it still does.

Decades of data collection from around the world has shown that there is a fair deal of consistency in what a baby's first words tend to be. The biggest data collection repository is the Wordbank,[9] an online database of tens of thousands of "communicative development inventories" that contains checklists of hundreds of nouns, verbs, adjectives, and pronouns that parents tick off if their children either say or understand them. In American English, for example, Wordbank shows that the ten most frequent first words, in order, are: *mommy, daddy, ball, bye, hi, no, dog, baby, woof woof,* and *banana.* In Hebrew, meanwhile, they are *mommy, yum yum, grandma, vroom, grandpa, daddy, banana, this, bye,* and *car.*[10] Chances are that one of these words will be your baby's first—after all, it is either an object that is directly in sight for the baby and/or a word that is said frequently in their presence.

Vocabulary growth is rapid after that first utterance, with infants learning tens of words a day, so that by the age of two, a "normal" vocabulary range can be between fifty to six hundred words, with a median of about three hundred.[11] In 2005, Deb Roy at the Massachusetts Institute of Technology's Media Lab began a fascinating experiment that involved recording almost every waking moment of his son's life from nine months to two years old.* He rigged his house with cameras and microphones to record all the spoken words that his son would have heard as well as his son's utterances, amounting to a corpus of eight million words.[12] At the start of the project, his son could say "mama." By the end of the experiment, the son had a vocabulary of around seven hundred words, and he also combined words so that his mean sentence length was 2.5 words. While simple words such as "cat" emerged first, as one would expect,

---

* Deb Roy gave a TED talk about his findings: www.ted.com/talks/deb_roy_the_birth_of_a_word.

many words that were tied to a location or activity were picked up sooner than more abstract words such as "the" or "and."

While individual or combinations of words are all well and good, the real fireworks in linguistic skills are not lit until around eighteen months. Syntax and grammar start to kick in, and rather than single words, babies begin to combine them to say things like "I sit," "Baby gone," and one particular favorite—"no bed." This basic understanding of grammar is a big moment for babies, one in which they show an amazing ability to use it correctly by putting words in the right place according to the rules of language. You will not hear a baby say, "Sit I" or "Go I," unless you gave birth to a rather green-looking baby with large ears and a remarkable ability to move objects with the "Force."

In the space of a year—from two to three years old—baby language becomes sophisticated incredibly quickly, with toddlers able to construct complex sentences. This development is so rapid that it is difficult for linguists to effectively study. I wish that I had written down what my two could say at different stages, but thankfully researchers have documented examples. Here is a boy called Adam and what he could say at two years and three months:*

"Big drum"; "I got horn"; "A bunny-rabbit walk."

Now compare that to what the same child could say four months later:

"Where piece of paper go?"; "Shadow has a hat like that"; "Dropped a rubber band."

And finally at three years old:

"You dress me up like a baby elephant"; "I going come in fourteen minutes"; "I going to wear that to wedding."

It is incredible to think that this small human who can only just about walk in a straight line is able to perform this level of neural gymnastics—a remarkable cognitive ability that sets us apart in the animal kingdom. Indeed, children learn their native language despite not having been formally taught grammatical rules. You only become aware that there are such formal

---

* Taken from: Pinker, S. *The Language Instinct.* Penguin Books (2015) 267–269.

constructions when you are taught the basics at school, even though you have been effectively using them since you were two years old.

The method by which infants acquire language and grammar is, again, a much-debated topic of current research among linguists. Some contest that humans are born with an innate knowledge of universal structural rules of grammar. This is suggested because some think that infants cannot possibly learn all the grammar rules by trial and error through incidental daily experience with language to then construct—and understand—sentences of such consistency, intricacy, and complexity.

But the idea has its critics who say it amounts to pseudoscience. Where there is little doubt is that infants have an amazing *capacity* to learn language incredibly fast. So, how is it possible for a child to go from saying a single word at twelve months old to creating grammatically correct sentences they have never heard just a year later?

Language is what makes human civilization possible. It is an innate part of human behavior and culture, so much so that there are around seven thousand different languages spoken around the world today.[13] When we speak, write, or talk, we use a relatively small number of words frequently, with the majority of possible words never used at all. As the eighteenth-century Prussian philosopher Wilhelm von Humboldt once noted, language makes "infinite use of finite means." In the one-million-word Brown Corpus of American English,* the top five words alone—*the*, *of*, *and*, *to*, *a*—account for 20 percent of usage. About 45 percent of words only appear once in a text or conversation. This "sparsity of language" is best shown by a rule named after George Kingsley Zipf, an American linguist who was born in Freeport, Illinois, in 1902.

Zipf's law[14] relates the frequency of an object to its ranking of use. It applies to many aspects of life—from the sizes of cities to that of galaxies in the cosmos. But its first use was in the study of language. Zipf's law states

---

* The Brown Corpus is a database of text samples of American English: korpus.uib.no/icame/manuals/BROWN/INDEX.HTM.

that the ranking of the usage of a word and its frequency of use are inversely related. In other words, in a corpus of written text such as the Bible or even this book, the second most frequently used word occurs approximately half as often as the first, and the third item a third as often as the first, and so on. The law holds for single words, but it also stands for two-word combinations. In this case, the effect is even more pronounced, with 80 percent of two-word combinations occurring only once and 90 percent of three-word combinations occurring only once. It's not clear whether there is any deeper meaning as to why many languages follow Zipf's law, but they do.

In the 1990s, the American linguist Noam Chomsky,* at the Massachusetts Institute of Technology, proposed "recursive" structures as the key element in human language.[15] "Recursive" means the repeated sequential use of a particular element or structure. This sounds strange but can be explained with this simple phrase: *Read the book*. Here, *the* and *book* are merged to produce (*the, book*), which is then merged with *read* to form (*read*, (*the, book*))—(*the, book*) being present in (*the, book*) and (*read*, (*the, book*)).[16] The basic premise of this merge is that, in all human languages, the relationships between words and the grammatical rules governing their combinations form a treelike network.

A full introduction to language theory is outside the scope of this book, but almost all human languages can be described as what is known as a context-free grammar.† Although it is important to note that a context-free grammar only covers syntax—how words fit together to make a sentence—and not semantics, which is the meaning of the sentence. According to Chomsky, syntax is a good starting point for a scientific approach to language. The definition of a context-free grammar, provided by Chomsky, is a set of (recursive) rules that generates a treelike structure. The three main aspects of a context-free grammar are "non-terminal" symbols, "terminal" symbols, and "production rules."

---

* Chomsky is also a proponent of "universal grammar," which contends that a certain set of structural grammatical rules are innate to humans.

† Except Swiss-German and Bambara. Some linguists also contest that context-free grammars are too simplified to describe language.

In a language, non-terminal symbols are things like noun phrases or verb phrases (i.e., parts of the sentence that can be broken down into smaller parts) while terminal symbols are produced when all the operations have been carried out, such as the individual words themselves. Finally, there are the hidden production rules that determine where the terminal symbols should be placed to produce a sentence that makes sense. A grammar is context-free if, after all the procedures are carried out, it all results in non-terminal symbols (or words).

A sentence in a context-free grammar language can be visualized like a tree with the branches being the "non-terminal" objects that the infant does not hear when learning language, such as verb phrases, and so on. The leaves of the tree, meanwhile, are the terminal symbols, or the actual words that are heard. For example, starting with the sentence "The baby vomited in the car" at the top of the tree, we can then split it into two: "The baby" being a noun phrase while "vomited in the car" is a verb phrase. But these parts are not the lowest denominator of the sentence—they can still be broken down. "The baby" can be broken into the determinant "the" and the noun "baby," while the verb phrase can be broken down with more steps (see figure).

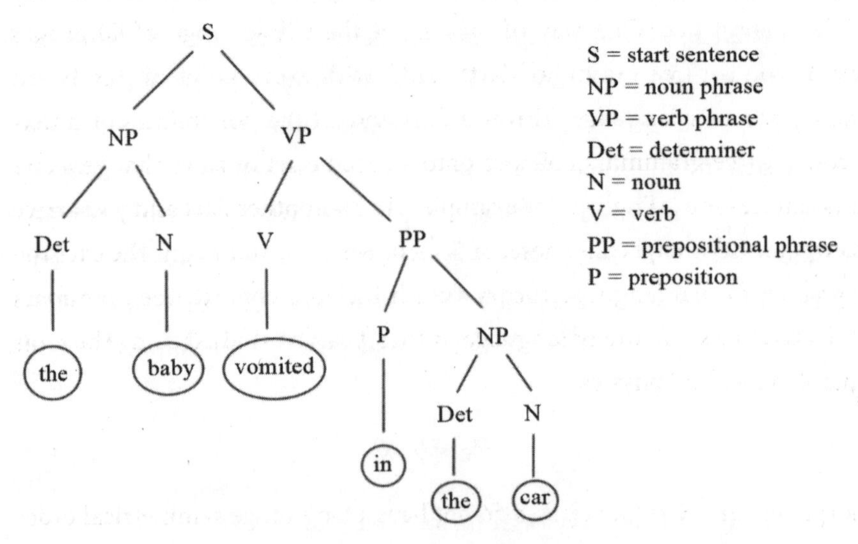

Fig 1 | *Breakdown of the sentence "The baby vomited in the car."*

Each of these subdivisions represents a branching point to reveal the final words, or "leaves," of the tree. When someone hears a sentence, this individual is mentally constructing the syntax tree for that sentence by building the correct "tree" that amounts to understanding the syntax of the sentence. For example, the sentence "There is a baby in a cot that can talk" is ambiguous. There are two different syntax trees corresponding to it, but only one of them is the correct one. The eureka moment you get from understanding language—according to this framework—is the moment of building the right tree.

A simple experiment in the 1980s elegantly described how language is hierarchical, or treelike, rather than understood in a simple word-by-word "linear" fashion. Children were shown a sequence of balls (red, green, green, red, red) and given the verbal instruction to point to "the second green ball." Following this instruction in a linear way would make you pick the second ball. It is second and green. But the children did not do this; instead, they picked the third ball, which was also green.[17] The reason for this is that they first merge the adjective "green" with the noun "ball" to understand that there is a set of green balls. This is then merged with "second" to pick up the second member of the set.

There are a whole host of context-free grammars, and they are not just limited to human languages. Computer languages such as Java and C++ are mainly context free. One way of looking at the whole range of languages is via a "context-free grammar dartboard," with every point on the board being a particular language. Throw a dart and all the possibilities of different context-free grammars collapse onto a specific set of rules that generate that language—be it English, for example. Throw another dart and you arrive somewhere else—maybe Japanese, or Java, or Russian, and so on. The exciting thing about formal language theory, which includes context-free grammars and the treelike structure of language, is that it can be studied using the tools of mathematics and physics.

Amorphous materials like glasses do not have a long-range symmetrical order as crystals do. Studying pure crystals can often be done by understanding the physics of their repetitive basic unit cell and extrapolating to the size of the sample. Yet, amorphous materials do not have this perfect self-repeating pattern, and so deducing their behavior often involves applying techniques gleaned from statistical physics to analyze the many smaller, variable interacting parts to see how they impact the whole structure. Theoretical physicist Eric De Giuli has been working in the field of amorphous solids since he earned his PhD in applied mathematics in 2013 from the University of British Columbia in Vancouver. When he later moved to the Philippe Meyer Institute of Theoretical Physics at the École Normale Supérieure in Paris as a postdoctoral researcher, he began to develop a theory to describe the specific states of amorphous solids.

While in Paris, De Giuli began to think about to what other topics he could apply the statistical techniques that he had been working on. One intriguing possibility that piqued his interest was language, especially given recent efforts by scientists to teach computers to understand human language, and so he began reading books to see how he could apply techniques from statistical physics to syntax. "I had total freedom to pursue what I wanted," De Giuli said, "and so I began to think what I could contribute to the theory of language using these methods."

When infants listen to people talking in fully formed sentences (that are, hopefully, grammatically correct), they are only exposed to the leaves, or surface, of the treelike network (the words and location in a sentence). But somehow, they also have to extract the rules of the language from the mixture of words they are hearing. De Giuli created a model in which the "surface" consists of all possible arrangements of words into sentences, including, in principle, nonsensical ones.[18] Attached to each branch going inside the structure of the tree is a "weight," or a probability of that particular sentence occurring. Initially, to a baby, all the weights are equal among all the nodes so that all possible sentence outcomes are equally likely. In this sense, the "language" is indistinguishable from random-word combinations and so carries with it no meaningful information.

As infants listen, they start to identify and learn individual words—the leaves of the tree—but cannot yet discern the deep branching structures underneath the leaves. Infants continually adjust the weights of the branches of possibilities as they hear language, so that eventually, branches that produce nonsensical sentences acquire smaller weights—because they are never heard—compared to information-rich branches that are given larger weights. By continuously performing this ritual of listening, the infant "prunes" the tree over time to discard random-word arrangements while retaining those with meaningful structure. This pruning process reduces both the number of branches near the tree's surface and those deeper down.

The fascinating aspect of this idea from a physical point of view is that when the weights are equal, language is random, which can be compared to how heat affects particles in thermodynamics. But once weights are added to the branches and adjusted to produce specific grammatical sentences, the "temperature" begins to decrease. When De Giuli ran his model for 25,000 possible distinct "languages" (which included computer languages), he found universal behavior when it came to "decreasing the temperature." At a certain point, there is a sharp drop in what is analogous to thermodynamic entropy, or disorder, when the language goes from a body of random arrangements to one that has high-information content.

Think of a bubbling pot of jumbled words that is taken off the stove to cool until words and phrases begin to "crystallize" into a specific structure or grammar. This abrupt switch is well known in statistical physics as a "phase transition"—an example in everyday life is the freezing of water to make ice cubes. As you reduce the temperature of the water, some parts of it—on the surface or the interior—start to freeze but the whole volume is not yet frozen. Then, suddenly, the whole cube freezes. In a similar way, at a certain point, the language switches from a random jumble of words to a highly structured communication system that is rich in information, containing sentences with complex structures and meanings.

De Giuli, who is now at Ryerson University in Canada, thinks that this model, which he stresses is only a model and not a definitive conclusion for how infants learn language, could explain why at a certain stage of development a child learns incredibly quickly to construct grammatical sentences.

There comes a point when they have listened to enough for it all to make sense to them. At that point, words cease to be mere labels and instead become the ingredients of sentences with complex structures and meanings. So, according to De Giuli's model, children are born with the ability to sample any context-free grammar language, and as they listen to their parents, or anyone else talking to them nearby, they zone into the specific rules—like the dart in our dartboard analogy—in whatever language they are learning. The only aspect that would be innate to humans in De Giuli's model is the ability to utilize this recursive treelike structure—everything else that emerges for the child to learn a language would come naturally from that.

This transition in real life may not happen so instantly, of course, but over time. And this also depends on getting all the correct weightings, something that children continue to refine as they grow. De Giuli says that the theory is consistent with psychologists' observations of language acquisition in young children. Research in 2009 seems to back up De Giuli's theory. Using data from the CHILDES corpus of first language acquisition,[19] researchers analyzed the words that a child named Peter spoke when he was around two years of age. From the various sentences spoken by Peter, the scientists constructed networks of the words that were used and analyzed the "hubs" of the network.[20] They found that at about two years of age, the hubs were simple words such as "it" that can mean multiple things. But just a few months later, the words began to carry syntactic meaning like "that," "a," "the," or "this." In fact, the language change at that age was so stark that it represented adult-like syntactic networks. Similar to what De Giuli found in his work, the research showed that the emergence of syntactic structure just after two years old is a sharp transition rather than something that is prolonged over many months.

According to De Giuli, there may be many such transitions occurring over time as children continue to learn and perfect their own language. De Giuli's theory could also apply to bilingualism, possibly resulting in two sets of weights being applied to the grammatical tree, although more work is needed in this regard. De Giuli also hopes that his results could help to inform neurological studies, potentially to learn what might inhibit the transition to rich language for children with learning disabilities. But, at the very least, it adds another idea to the many growing theories of how grammar can develop in children.

For many parents, the ability to communicate with their children is a major milestone. Of course, it still will not stop that meltdown in the middle of the supermarket because they could not have the yogurt that features their favorite cartoon character. But just being able to converse makes things so much easier. The spark of language also represents the start of their detachment, ever so slightly, from you as a parent. Now, they can talk to other people—tell them what is on their minds, what they want, what they need—in forming new relationships. Language is the main tool they have at their disposal that will help to guide them in the world as they explore it for themselves, make friends, discover new interests, and ultimately discover who they are.

AFTERWORD

When I first began working on this book, I thought it would feature mostly amusing stories and anecdotes of how physicists, mathematicians, and engineers have attempted, and possibly failed, to study aspects of the physics of babies. After all, you can usually find the physics of "anything" if you look hard enough. I had in mind that a lot of the research I would include would be historical, discovering the odd research paper done twenty or thirty years ago and nothing since—perhaps being someone's side project away from their "serious" research. But the more research I did, the more I found that not only are these lines of research far from a bit on the side, but many of the topics covered in the book, if not all, are active areas of research—most of which have just been done in the past decade.

Pregnancy is not a rare circumstance—a significant fraction of women become pregnant at least once in their lifetime. Yet for too long, pregnancy has been seen as something to be endured rather than studied. And such research, especially when it comes to a physical understanding of pregnancy and childbirth, has lagged woefully behind that of other areas in physical biology. Conception from a physical-science perspective is somewhat of a black box. We still do not understand how contractions start or propagate throughout the uterus or how the placenta manages to control the diffusion of a wide range of solutes. We know far more about cow's milk than we do about human milk, and we do not have a physical explanation of conditions such as sudden infant death syndrome. The placenta, uterus, and cervix from a physical perspective are drastically understudied—they are decades behind work on other organs such as the heart, kidney, liver, lungs, and brain, which often have objective function tests to see in detail how the organ is working. Some of the topics and research we have come across in this book have been done by scientists who were surprised—possibly shocked—by how little we know about a particular area of pregnancy.

It begs the question: Why does a physical understanding of, say, the placenta or birth lag so far behind? For several reasons. One is the difficulty from an ethical perspective of studying pregnant women and newborns. Studies must be carefully designed, and various ethical procedures and guidelines need to be adhered to—and quite rightly too. The other is more disturbing and can likely be laid at the historic dominance of men in engineering, mathematics, and physics, some of whom have failed to see the interest or importance of carrying out research on such issues. You might ask how today we have such an advanced knowledge of the heart and its function—is it by any chance because heart disease is a leading cause of death in men? For a male-dominated funding panel, it would be easy to see why they would allocate research funds into heart disease—something that is likely to affect them soon enough—instead of examining the basics of the placenta or the uterus.

In 2018, the *Guardian* newspaper in the UK reported that 90 percent of funding awarded by the Engineering and Physical Sciences Research Council—one of the main research agencies in the UK—went to projects led by men.[1] The gender gap in science, and especially physics, is well documented, with women accounting for about 20 percent at the undergraduate level—that percentage dropping further up the career ladder. Things may be beginning to change, slowly, with women taking up more senior positions and sitting on funding panels, but there is still some way to go on the equality side.

Of course, it would be crass to think that physics, mathematics, or engineering can solve every problem to do with conception and pregnancy, but they can certainly help to bring a different—and welcome—perspective. The power of the physical sciences comes from its tools and theoretical techniques. Yet applying those to areas outside the traditional scope of the subject cannot be done in a siloed way. It is most effective when scientists work with experts in that area—be they fertility specialists, pediatricians, gynecologists, midwives, physiologists, or neuroscientists. Many of the mathematicians, engineers and physicists we have met in this book have pioneered just that—sometimes having to themselves learn a new scientific "language" in the process.

Only with more funding and support for such multidisciplinary collaborations will we get a clearer view of conception, pregnancy, childbirth, and infant development—work that promises new insights and may also lead to new clinical treatments and applications.

After all, such developments are not only important for now but for the generations to come.

ACKNOWLEDGMENTS

This book would not have been possible without the help of so many people who kindly gave their time to talk about their research. A special mention goes to Dave Smith, who organized and hosted a visit to Birmingham, UK, to learn more about his work on sperm mechanics and embryology; likewise to Hermes Bloomfield-Gadêlha, who gave me an afternoon one day in Bristol to talk about swimming sperm. Sadly, the COVID-19 pandemic and subsequent lockdowns closed the door on further visits, but thankfully video conference tools came to the fore, and many interviews with experts were conducted in this way.

I have always found scientists to not only be incredibly enthusiastic about their work but also generous with their time, and the people I interviewed for this book did not disappoint. The pandemic made time even more precious, so I can only thank them for squeezing me in to their schedules when so much about normal life was upside down. In no particular order: Ida Sabelis, Jackson Kirkman-Brown, Jean-Léon Maître, Otger Campàs, Niamh Nowlan, Robert Catena, Igor Chernyavsky, Alys Clark, Penny Gowland. Megan Leftwich, Nicolas Garnier, Jennifer Kruger, Andrew Blanks, Mehdi Raessi, Cory Hoi, Ravi Selvaganapathy, Ingo Titze, Hanspeter Herzel, David Elad, Donna Geddes, Michael Woolridge, Ronny Bartsh, Natalie Schaworonkow, Francesca Sylos-Labini, Marianne Barbu-Roth, David Anderson, and Eric DeGuili. Despite their help to make the book scientifically accurate and error-free, any mistakes that remain in the text are all mine.

I would also like to thank Glenn Yeffeth at BenBella Books for seeing the potential in a kernel of an idea that has now, hopefully, blossomed into an interesting and enjoyable book. Thank you to Alexa Stevenson and Jodi Frank for their excellent edits on the draft manuscript, improving it vastly, as well as to everyone at BenBella Books for their work turning a concept and words on a page into a full-fledged book. Thank you to Matin Durrani for advice as I veered into the world of book publishing, as well as to Michael Allen and Marric Stephens for reading through early draft chapters and offering their comments, suggestions, and corrections.

Finally, to my family. My mother, who, after my father died suddenly when I was entering my teenage years, supported my interests and created an environment where I could follow my ambitions—*per aspera ad astra.* I also want to thank Claire, especially for supporting this project and allowing me the time to focus and work on the book despite the need to bring up two young, energetic boys. Thank you. And to Henry and Elliott. This book would never have existed if you both did not enter our lives. We are blessed that you have done so, and you will always brighten the darkest days. Stay curious; it will lead you to where you need to go.

ENDNOTES

CHAPTER 1

1. Sabelis, I. "To Make Love as a Testee." *Annals of Improbable Research* 7, no.1 (2001): 14–15.
2. Schultz, W.W., van Andel, P., Sabelis, I., et al. "Magnetic Resonance Imaging of Male and Female Genitals During Coitus and Female Sexual Arousal." *British Medical Journal* 319 (1999):1596–1600.

CHAPTER 2

1. Retrieved from https://www.pepysdiary.com/diary/1665/01/21/.
2. van Zuylen, J. "The Microscopes of Antoni van Leeuwenhoek." *Journal of Microscopy* 121 (1981): 309–328.
3. Cocquyt, T., Zhou, Z., Plomp, J., et al. "Neutron Tomography of van Leeuwenhoek's Microscopes." *Science Advances* 7 (2021): eabf2402.
4. Gest, H. "The Discovery of Microorganisms by Robert Hooke and Antoni van Leeuwenhoek, Fellows of the Royal Society." *Notes and Records: The Royal Society Journal of the History of Science* 58, no.2 (2004): 187–201.
5. Cobb, M. "An Amazing 10 Years: The Discovery of Egg and Sperm in the 17th Century." *Reproduction in Domestic Animals* 47 (2012): 2–6.
6. Retrieved from https://www.nationalgeographic.com/science/article/100318-men-sperm-1500-stem-cells-second-male-birth-control.
7. Boskey, E.R., Cone, R.A., Whaley, K.J., et al. "Origins of Vaginal Acidity: High D/L Lactate Ratio Is Consistent with Bacteria Being the Primary Source." *Human Reproduction* 16 (2001): 1809–1813.
8. Retrieved from https://www.flagellarcapture.com/learn/background.

9. Fukuda, M., and Fukuda, K. "Uterine Endometrial Cavity and Movement and Cervical Mucus." *Human Reproduction* 9 (1994):1013–1016.
10. Reynolds, O. "An Experimental Investigation of the Circumstances Which Determine Whether the Motion of Water Shall Be Direct or Sinuous and of the Law of Resistance in Parallel Channels." *Philosophical Transactions of the Royal Society* 174 (1883): 935–982.
11. Klotsa, D. "As Above, So Below, and Also in Between: Mesoscale Active Matter in Fluids." *Soft Matter* 15 (2019): 8946–8950.
12. Bleaney, B. "Edward Mills Purcell. 30 August 1912–7 March 1997." *Biographical Memoirs of Fellows of the Royal Society* 45 (1999): 437–447.
13. Purcell, E.M. "Life at Low Reynolds Number." *American Journal of Physics* 45 (1977): 3–11.
14. Retrieved from http://web.mit.edu/hml/ncfmf/07LRNF.pdf.
15. Chwang, A.T., and Wu, T.Y. "A Note on the Helical Movement of Mirco-Organisms." *Proceedings of the Royal Society of London B* 178, no.1052 (1971): 327–346.
16. Taylor, G. "Analysis of the Swimming of Microscopic Organisms" *Proceedings of the Royal Society of London A* 209 (1951): 447–461; and Hancock, G.J. "The Self-Propulsion of Microscopic Organisms Through Liquids." *Proceedings of the Royal Society A* 217 (1953): 96–121.
17. Gray, J., and Hancock, G.J. "The Propulsion of Sea-Urchin Spermatozoa." *Journal of Experimental Biology* 32, no.4 (1955): 8032–8114.
18. Afzelius, B. "Electron Microscopy of the Sperm Tail; Results Obtained with a New Fixative." *Journal of Biophysical and Biochemical Cytology* 5 (1959): 269–278.
19. Gaffney, E.A., Gadêlha, H., Smith, D.J., et al. "Mammalian Sperm Motility: Observation and Theory." *Annual Review of Fluid Mechanics* 43 (2011): 501–528.
20. Lindemann, C.B., Macauley, L.J., and Lesich, K.A. "The Counterbend Phenomenon in Dynein-Disables Rat Sperm Flagella and What It Reveals About the Interdoublet Elasticity." *Biophysical Journal* 89, no.2 (2005): 1165–1174.
21. Ishimoto, K., Gadêlha, H., Gaffney, E.A., et al. "Coarse-Graining the Fluid Flow Around a Human Sperm." *Physical Review Letters* 118 (2017): 124501–124505.
22. Gadêlha, H., and Gaffney, E.A. "Flagellar Ultrastructure Suppresses Buckling Instabilities and Enables Mammalian Sperm Navigation in High-Viscosity Media." *Journal of Royal Society Interface* 16 (2019): 20180668.
23. Corkidi, G., Hernández-Herrera, P., Montoya, F., et al. "Long-Term Segmentation-Free Assessment of Head-Flagellum Movement and Intracellular Calcium in Swimming Human Sperm." *Journal of Cell Science* 134, no.3 (2021): jcs250654.
24. Suarez, S.S., and Pacey, A.A. "Sperm Transport in the Female Reproductive Tract." *Human Reproduction Update* 12 (2005): 23–37.
25. Fitzpatrick, J.L., Willis, C., Devigili, A., et al. "Chemical Signal from Eggs Facilitate Cryptic Female Choice in Humans." *Proceedings of the Royal Society B* 287 (2020): 20200805.

26. Stünker, T., Goodwin, N., Brenker, C., et al. "The CatSper Channel Mediates Progesterone-Induced Ca2+ Influx in Human Sperm." *Nature* 471 (2011): 382–386; and Lishko, P. V., Botchkina, I.L., Kirichok, Y. "Progesterone Activates the Principal Ca2+ Channel of Human Sperm." *Nature* 471 (2011): 387–391.
27. Ded, L., Hwang, J.Y., Miki, K., et al. "3D In Situ Imaging of the Female Reproductive Tract Reveals Molecular Signatures of Fertilizing Spermatozoa in Mice." *eLife* 9 (2020): e62043.
28. Human Fertilisation and Embryology Authority, *Fertility Treatment 2017: Trends and Figures* (2019): https://www.hfea.gov.uk/media/2894/fertility-treatment-2017-trends-and-figures-may-2019.pdf.
29. Human Fertilisation and Embryology Authority, *Fertility Treatment 2014–16: Trends and Figures* (2018).
30. Cohen, J., and McNaughton, D.C "Spermatozoa: The Probable Selection of a Small Population by the Genital Tract of the Female Rabbit." *Reproduction* 39, no.2 (1974): 297–310.
31. Gallagher, M.T., Smith, D.J., and Kirkman-Brown, J.C. "CASA: Tracking the Past and Plotting the Future." *Reproduction, Fertility and Development* 30 (2018): 867–874.
32. Gallagher, M.T., Cupples, G., Ooi, E.H., et al. "Rapid Sperm Capture: High-Throughput Flagellar Waveform Analysis." *Human Reproduction* 34, no.7 (2019):1173–1185.
33. Elad, D., Jaffa, A., and Grisaru, D. "Biomechanics of Early Life in the Female Reproductive System." *Physiology* 35 (2020): 134–143.

CHAPTER 3

1. D'Arcy, W.T. *On Growth and Form.* Cambridge University Press (1917), Cambridge.
2. Jarron, M. "Cell and Tissue, Shell and Bone, Leaf and Flower—On Growth and Form in Context." *Mechanisms of Development* 145 (2017):22–25.
3. Turing, A. "The Chemical Basis of Morphogenesis." *Philosophical Transactions of the Royal Society of London B 237 (1952): 37–72.*
4. Murray, J. "How the Leopard Gets Its Spots." *Scientific American* 258, no.3 (1988): 80–87.
5. Sick, S., Reinker, S., Timmer, J., et al. "WNT and DKK Determine Hair Follicle Spacing Through a Reaction-Diffusion Mechanism." *Science* 314 (2006): 1447–1450.
6. Sheth, R., Marcon, L., Bastida, M.F., et al. "Hox Genes Regulate Digit Patterning by Controlling the Wavelength of a Turing-Type Mechanism." *Science* 338 (2012): 1476–1480.
7. Raspopovic, J., Marcon, L., Russo, L., et al. "Digit Patterning Is Controlled by a Bmp-Sox9-Wnt Turing Network Modulated by Morphogen Gradients." *Science* 345 (2014): 566–570.

8. Duncan, F.E., Que, E.L., Zhang, N., et al. "The Zinc Spark Is an Inorganic Signature of Human Egg Activation." *Scientific Reports* 6 (2016): 24737. doi:10.1038/srep24737.
9. Dumortier, J.G., Le Verge-Serandour, M., Tortorelli, A.F., et al. "Hydraulic Fracturing and Active Coarsening Position the Lumen of the Mouse Blastocyst." *Science* 365 (2019): 465–468.
10. Baillie, M. "An Account of a Remarkable Transposition of the Viscera." *Philosophical Transactions of the Royal Society of London* 78 (1788): 350–363.
11. Nonaka, S., Shiratori, H., Saijoh, Y., et al. "Determination of Left-Right Patterning of the Mouse Embryo by Artificial Nodal Flow." *Nature* 418 (2002): 96–99.
12. Cartwright, J.H.E., Piro, O., and Tuval, I. "Fluid-Dynamical Basis of the Embryonic Development of Left-Right Asymmetry in Vertebrates." *Proceedings of the National Academies of Sciences* 101 (2004): 7234–7239.
13. Nonaka, S., Yoshiba, S., Watanabe D., et al. "De Novo Formation of Left-Right Asymmetry by Posterior Tilt of Nodal Cilia." *PLOS Biology* 3, no.8 (2005): e268. doi:10.1371/journal.pbio.0030268.
14. Okada, Y., Takeda, S., Tanaka Y., et al. "Mechanism of Nodal Flow: A Conserved Symmetry Breaking Event in Left-Right Axis Determination." *Cell* 121 (2005): 633–644.
15. Smith, D.J., Montenegro-Johnson, T.D., and Lopes, S.S. "Symmetry-Breaking Cilia-Driven Flow in Embryogenesis." *Annual Review of Fluid Mechanics* 51 (2019): 105–128.
16. Mongera, A., Rowghanian, P., Gustafson, H.J., et al. "A Fluid-to-Solid Jamming Transition Underlies Vertebrate Body Axis Elongation" *Nature* 561 (2018):401–405.

CHAPTER 4

1. Barrett, J.C., and Marshall, J. "The Risk of Conception on Different Days of the Menstrual Cycle." *Population Studies* 23 (1969): 455–461.
2. Schwartz, D., Macdonald, P.D.M., and Heuchel, V. "Fecundability, Coital Frequency and the Viability of Ova." *Population Studies* 34, no.2 (1980): 397–400.
3. Wilcox, A.J., Dunson, D.B., and Baird, D.D. "The Timing of the 'Fertile Window' in the Menstrual Cycle: Day Specific Estimates from a Prospective Study." *British Medical Journal* 321 (2000): 1259–1262.
4. Wilcox, A.J., Baird, D.D., Dunson, D.B., et al. "On the Frequency of Intercourse Around Ovulation: Evidence for Biological Influences." *Human Reproduction* 19, no.7 (2004): 1539–1543.
5. Scarpa, B., Dunson, D.B., and Giacchi, E. "Bayesian Selection of Optimal Rules for Timing Intercourse to Conceive by Using Calendar and Mucus." *Fertility and Sterility* 88, no.4 (2007): 915–924.
6. Dunson, D.B., Baird, D.D., and Columbo, B. "Increased Infertility with Age in Men and Women." *Age and Infertility* 103, no.1 (2004): 51–56.
7. Retrieved from https://www.naturalcycles.com/.

8. Bull, J.R., Rowland, S.P., Berglund Scherwitzl, E., et al. "Real-World Menstrual Cycle Characteristics of More Than 600 000 Menstrual Cycles." *NPG Digital Medicine* 83 (2019). doi.org/10.1038/s41746-019-0152-7.
9. Ghalioungui, P., Khalil, S.H., and Ammar, A.R. "On an Ancient Egyptian Method of Diagnosing Foetal Sex." *Medical History* 7, no.3 (1963): 241–246.
10. Herbert, E., and Simpson, M. "Aschheim-Zondek Test for Pregnancy—Its Present Status." *California and Western Medicine* 32 (1930):145–148.
11. Howe, M. "Dr. Maurice Friedman, 87, Dies: Created Rabbit Pregnancy Test." *New York Times*, March 10, 1991.
12. Well, G.R. "Lancelot Thomas Hogben, 9 December 1985–22 August 1975." *Biographical Memoirs of Fellows of the Royal Society* 24 (1978): 183–221.
13. Science Museum. Retrieved from http://broughttolife.sciencemuseum.org.uk/broughttolife/techniques/frogs.
14. Vredenburg, V.T., Felt, S.A., Morgan, E.C., et al. "Prevalence of *Batrachochytrium dendrobatidis* in *Xenopus* Collected in Africa (1871–2000) and in California (2001–2010)." *PLOS One* 8, no.5 (2013): e63791. doi.10.1371/journal.pone.0063791.
15. Wide, L. "Inventions Leading to the Development of the Diagnostic Test Kit Industry—From the Modern Pregnancy Test to the Sandwich Assay." *Upsala Journal of Medical Sciences* 110, no.3 (2005): 193–216.
16. Swaminathan, N., and Bahl, Om.P. "Dissociation and Recombination of the Subunits of Human Chorionic Gonadotropin." *Biochemical and Biophysical Research Communications* 40, no.2 (1970): 422–427.
17. Cowsill, B.J. "The Physics of Pregnancy Tests: A Biophysical Study of Interfacial Protein Adsorption." PhD thesis. UK: University of Manchester, 2012.

FIRST INTERLUDE

1. Donald, I., MacVicar, J., and Brown, T.G. "Investigation of Abdominal Masses by Pulsed Ultrasound." *Lancet* 1 (1958): 1188–1195.
2. Campbell, S. "A Short History of Sonography in Obstetrics and Gynaecology." *Facts, Views and Vision in ObGyn* 5 (2013): 213–229.
3. Jouppila, P. "Ultrasound in the Diagnosis of Early Pregnancy and Its Complications: A Comparative Study of the A-, B- and Doppler Methods." *Acta Obstetricia et Gynecologica Scandinavica* 50, no S15 (1971): 1–56.
4. Retrieved from https://mathshistory.st-andrews.ac.uk/Biographies/Doppler/.
5. Buys Ballot, C. H. Dus. "Akustische Versuche auf der Niederländische Eisenbahn, nebst gelegentliche Bemerkungen zur Theorie des Herrn Prof. Doppler." *Annalen der Physik und Chemie* 66 (1845): 321–351.
6. Mauli, D. "Doppler Sonography: A Brief History." *Doppler Ultrasound in Obstetrics and Gynaecology*, Springer (2005) Berlin.
7. Johnson, W.L., Stegall, H.F., Lein, J., et al. "Detection of Fetal Life in Early Pregnancy with an Ultrasonic Doppler Flowmeter." *Obstetrics and Gynecology* 26, no.3 (1965): 305–306.

8. Fitzgerald, D.E., and Drumm, J.E. "Non-invasive Measurement of Human Fetal Circulation Using Ultrasound: a New Method." *British Medical Journal* 2 (1977): 1450–1451.
9. Retrieved from https://www.fda.gov/consumers/consumer-updates/avoid-fetal-keepsake-images-heartbeat-monitors.
10. Retrieved from https://www.isuog.org/clinical-resources/isuog-guidelines/practice-guidelines-english.html.

CHAPTER 5

1. De Vries, J.I.P., and Fong, B.F. "Normal Fetal Motility: An Overview." *Ultrasound Obstetrics & Gynecology* 27, no.6 (2006): 701–711.
2. Whitehead, K., Meek, J., and Fabrizi, L. "Developmental Trajectory of Movement-Related Cortical Oscillations During Active Sleep in a Cross-sectional Cohort of Pre-term and Full-term Human Infants." *Scientific Reports* 8 (2018): 17516.
3. Dutton, P.J., Warrander, L.K., Roberts, S.A., et al. "Predicators of Poor Perinatal Outcome Following Maternal Perception of Reduced Fetal Movements—a Prospective Cohort Study." *PLoS One* 7 (2012): e39784.
4. Hertogs, K., Roberts, A.B., Cooper, D., et al. "Maternal Perception of Fetal Motor Activity." *British Medical Journal* 2 (1979): 1183–1185.
5. Verbruggen, S.W., Kainz, B., Shelmerdine, S.C., et al. "Stresses and Strains on the Human Fetal Skeleton During Development." *Journal of the Royal Society: Interface* 15 (2018): 20170593. doi:10.1098/rsif.2017.0593.
6. Lai, J., Woodward, R., Alexandrov, Y., et al. "Performance of a Wearable Acoustic System for Fetal Movement Discrimination." *PLoS One* 13, no.5 (2018): e0195728. doi: 10.1371/journal.pone.0195728.
7. Thurber, C., Dugas, L.R., Ocobock, C., et al. "Extreme Events Reveal an Alimentary Limit on Sustained Maximal Human Energy Expenditure." *Science Advances* 5 (2019): eaaw0341. doi: 10.1126/sciadv.aaw0341.
8. Institute of Medicine and National Research Council. W*eight Gain During Pregnancy: Reexamining the Guidelines.* The National Academies Press (2009) Washington, DC: https://doi.org/10.17226/12584.
9. Kuo, C., Jamieson, D.J., McPheeter M.L., et al. "Injury Hospitalizations of Pregnant Women in the United States, 2002." *American Journal of Obstetrics and Gynecology* 196, no.2 (2007): 161.e1-161.e6.
10. Whitcome, K.K., Shapiro, L.J., and Lieberman, D.E. "Fetal Load and the Evolution of Lumbar Lordis in Bipedal Hominins." *Nature* 450 (2007): 1075–1078.
11. Branco, M., Santos-Rocha, R., and Filomena, V. "Biomechanics of Gait During Pregnancy." *Scientific World Journal* 2014 (2014): 527940.
12. Catena, R.D., Connolly, C.P., McGeorge, K.M., et al. "A Comparison of Methods to Determine Center of Mass During Pregnancy." *Journal of Biomechanics* 71 (2018): 217–224.
13. Dunning, K., LeMasters, G., and Bhattacharya, A. "A Major Public Health Issue: The High Incidence of Falls During Pregnancy." *Journal of Maternal and Child Health* 14 (2010): 720–725.

14. Catena, R.D., Campbell, N., Werner, A.L., et al. "Anthropometric Changes During Pregnancy Provide Little Explanation of Dynamic Balance Changes." *Journal of Applied Biomechanics* 35 (2019):232–239.
15. Catena, R.D., and Wolcott, W.C. "Self-selection of Gestational Lumbopelvic Posture and Bipedal Evolution." *Gait & Posture* 89 (2021): 7–13.

CHAPTER 6

1. Gunn, G.C., Mishell Jr., D.R., and Morton, D.G. "Premature Rupture of the Fetal Membranes—A Review." *American Journal of Obstetrics and Gynecology* 106, no.3 (1970): 469–483.
2. Young, R.C. "Myocytes, Myometrium and Uterine Contractions." *Annals of the New York Academy of Sciences* 1101 (2007): 72–84.
3. Kavanagh, J., Kelly, A.J., and Thomas, J. "Breast Stimulation for Cervical Ripening and Induction of Labour." *Cochrane Database of Systematic Reviews* 3 (2005): article number CD003392. doi: 10.1002/14651858.CD003392.pub2.
4. Retrieved from https://www.cdc.gov/reproductivehealth/features/premature-birth/index.html.
5. Retrieved from https://www.who.int/news-room/fact-sheets/detail/preterm-birth.
6. Tong, S., Kaur, A., Walker, S.P., et al. "Miscarriage Risk for Asymptomatic Women After a Normal First-Trimester Prenatal Visit." *Obstetrics and Gynecology* 111, no.3 (2008):710–714.
7. Keith, A., and Flack, M. "The Form and Nature of the Muscular Connections Between the Primary Divisions of the Vertebrate Heart." *Journal of Anatomy and Physiology* 41 (1907); 172–189.
8. Popescu, L.M., Ciontea, S.M., and Cretolu, D. "Interstitial Cajal-like Cells in the Human Uterus and Fallopian Tube." *Annals of the New York Academy of Science* 1101 (2005): 139–165.
9. Schwiening, C.J. "A Brief Historical Perspective: Hodgkin and Huxley." *Journal of Physiology* 590 (2012): 2571–2575.
10. Hodgkin, A.L., and Huxley, A.F. "A Quantitative Description of Membrane Current and Its Application to Conduction and Excitation in Nerve." *Journal of Physiology* 117 (1952): 500–544.
11. Fitzhugh, R. "Impulses and Physiological States in Theoretical Models of Nerve Membrane." *Biophysical Journal* 1, no.6 (1961): 445–466.
12. Keener, J., and Sneyd, J. "Mathematical Physiology," Second Edition. Springer (2009), New York.
13. Parkington, H.C., Tonta, M.A., Brennecke, S.P., et al. "Contractile Activity, Membrane Potential, and Cytoplasmic Calcium in Human Uterine Smooth Muscle in the Third Trimester of Pregnancy and During Labor." *American Journal of Obstetrics and Gynecology* 181, no. 6 (1999): 1445–1451.
14. Miyoshi, H., Boyle, M.B., MacKay, L.B., et al. "Voltage-Clamp Studies of Gap Junctions Between Uterine Muscle Cells During Term and Preterm Labor." *Biophysical Journal* 71 (1996): 1324–1334.

15. Singh, R., Xu, J., Garnier, N., et al. "Self-Organized Transition to Coherent Activity in Disordered Media." *Physical Review Letters* 108, no.6 (2012): 068102.
16. Xu, J., Menon, S.N., Singh, R., et al. "The Role of Cellular Coupling in the Spontaneous Generation of Electrical Activity in Uterine Tissue." *PLoS ONE* 10, no 3 (2015) e0118443.
17. Lammers, W.J.E.P., Mirghani, H., Stephen, B., et al. "Patterns of Electrical Propagation in the Intact Pregnant Guinea Pig Uterus." *American Journal of Physiology—Regulatory, Integrative and Comparative Physiology* 294, no.3 (2008): R919–R928.
18. Ghosh, R., Seenivasan, P., Menon, S.N., et al. "Frequency Gradient in Heterogeneous Oscillatory Media Can Spatially Localize Self-Organized Wave Sources That Coordinate System-Wide Activity." arXiv:1912.07271.
19. Sparey, C., Robson, S.C., Bailey, J., et al. "The Differential Expression of Myometrial Connexin-43 Cyclooxygenase-1 and -2, and Gsα Proteins in the Upper and Lower Segments of the Human Uterus During Pregnancy and Labor." *Journal of Clinical Endocrinology & Metabolism* 84, no. 5 (1999): 1705–1710.
20. Lutton, E.J., Lammers, W.J.E.P., James, S., et al. "Identification of Uterine Pacemaker Regions at the Myometrial-Placental Interface in Rats." *Journal of Physiology* 14 (2018): 2841–2852.

CHAPTER 7

1. Retrieved from https://patents.google.com/patent/US3216423A/en.
2. Retrieved from https://dublin.sciencegallery.com/fail-better-exhibits/apparatus-for-facilitating-the-birth-of-a-child-by-centrifugal-force.
3. World Health Organization. "Maternal Mortality: Levels and Trends 2000 to 2017." ISBN: 978-92-4-151648-8 (2019).
4. Boerma, T., Ronsmans, C., Meless, D.Y., et al. "Global Epidemiology of Use of and Disparities in Caesarean Sections." *Lancet* 392, no.10155 (2018): 1341–1348.
5. World Health Organization. Global Health Organization Data Repository: https://apps.who.int/gho/data/node.main.BIRTHSBYCAESAREAN?lang=en.
6. Hoxha I., Syrogiannouli, L., Braha M., et al. "Caesarean Sections and Private Insurance: Systematic Review and Meta-analysis." *BMJ Open* 7 (2017): e016600.
7. Retrieved from https://www.theatlantic.com/ideas/archive/2019/10/c-section-rate-high/600172/.
8. Dominguez-Bello, M.G., Costello, E.K., and Contreras, M. "Delivery Mode Shapes the Acquisition and Structure of the Initial Microbiota Across Multiple Body Habitats in Newborns." *Proceedings of the National Academy of Sciences* 107 (2010): 11971–11975.
9. Stokholm, J., Thorsen, J., Rasmussen, M.A., et al. "Delivery Mode and Gut Microbial Changes Correlate with an Increased Risk of Childhood Asthma." *Science Translational Medicine* 12 (2020): eaax9929.

10. Dominguez-Bello, M.G., De Jesus-Laboy, K.M., Shen N., et al. "Partial Restoration of the Microbiota of Cesarean-Born Infants via Vaginal Microbial Transfer." *Nature Medicine* 22 (2016): 250–253.
11. Korpela, K., Helve, O., Kolho, K-L., et al. "Maternal Fecal Microbiota Transplantation in Cesarean-Born Infants Rapidly Restores Normal Gut Microbial Development: A Proof-of-Concept Study." *Cell* 183 (2020): 1–11.
12. Retrieved from https://www.ouh.nhs.uk/patient-guide/leaflets/files/12101Ptear.pdf.
13. Rortveir, G., Daltveit, A.K., Hannestad, Y.S., et al. "Urinary Incontinence After Vaginal Delivery or Caesarean Section." *New England Journal of Medicine* 348 (2003): 990–997.
14. Eason, E., Labrecque, M., Marcoux, S., et al. "Anal Incontinence After Childbirth." *Canadian Medical Association Journal* 166, no.3 (2002): 326–330.
15. Dietz, H.P., and Simpson, J.M. "Levator Trauma Is Associated with Pelvic Organ Prolapse." *An International Journal of Obstetrics and Gynaecology* 115 (2008): 979–984.
16. Retrieved from https://americanpregnancy.org/healthy-pregnancy/labor-and-birth/second-stage-of-labor-897/.
17. Retrieved from https://www.royalberkshire.nhs.uk/patient-information-leaflets/Maternity/Maternity---shoulder-dystocia.htm.
18. Caldwell, W.E., and Moloy, H.C. "Anatomical Variations in the Female Pelvis: Their Classification and Obstetrical Significance." *Proceedings of the Royal Society of Medicine* 32, no.1 (1938): 1–30.
19. Betti, L., and Manica, A. "Human Variation in the Shape of the Birth Canal Is Significant and Geographically Structured." *Proceedings of the Royal Society B* 285 (2018): 20181807. doi: 0.1098/rspb.2018.1807.
20. Retrieved from https://www.theguardian.com/science/2018/oct/24/focus-on-western-women-skewed-our-ideas-of-what-birth-should-look-like.
21. Jing, D., Lien, K., Ashton-Miller, J.A., et al. "Visco-hyperelastic Properties of the Pelvic Floor Muscles in Healthy Women." *Proceedings of the North American Congress on Biomechanics*, Ann Arbor, Michigan, US: http://ww.asbweb.org/conferences/2008/abstracts/562.pdf.
22. Yan X., Kruger, J.A., Nielsen P.M.F., et al. "Effects of Fetal Head Shape Variation on the Second Stage of Labour." *Journal of Biomechanics* 48 (2015): 1593–1599.
23. Yan, X., Kruger, J.A., Li, X., et al. "Modeling the Second Stage of Labour." *WIREs Systems Biology and Medicine* 8 (2016): 506–516.
24. Noel, A., Guo, H., Mandica M., et al. "Frogs Use a Viscoelastic Tongue and Non-Newtonian Saliva to Catch Prey." *Journal of the Royal Society Interface* 14 (2017) 0160764. doi:10.1098/rsif.2016.0764.
25. Lehn, A.M., Baumer, A., Leftwich, M.C. "An Experimental Approach to a Simplified Model of Human Birth." *Journal of Biomechanics* 49 (2016): 2313–2317.
26. Golay, J., Vedam, S., and Sorger, L. "The Squatting Position for the Second Stage of Labor: Effects on Labor and on Maternal and Fetal Well Being." *Birth* 20 (1993): 73–78.

27. Varney, H., Kriebs, J.M., and Gegor, C.L. *Varney's Midwifery*, Jones and Bartlett Publishers (2004), Sudbury, MA.

SECOND INTERLUDE

1. Krafchik, B. "History of Diaper and Diapering." *International Journal of Dermatology* 55 (2016): 4–6.
2. Dyer, D. "Seven Decades of Disposable Diapers: A Record of Continuous Innovation and Expanding Benefit." European Disposables and Nonwovens Association. Brussels, Belgium (2005).
3. Retrieved from https://www.businesswire.com/news/home/20211203005510/en/Global-Baby-Diapers-Market-Report-2021-Market-is-Expected-to-Reach-65.50-Billion-in-2025-at-a-CAGR-of-6.8.
4. Retrieved from https://baby.lovetoknow.com/baby-care/how-many-diapers-does-baby-use-year.
5. Retrieved from https://bbia.org.uk/wp-content/uploads/2020/11/A-Circular-Economy-for-Nappies-final-oct-2020.pdf.
6. Retrieved from https://wrap.org.uk/resources/guide/waste-prevention-activities/real-nappies/overview.
7. "Advances in Technical Nonwovens." Edited by George Kellie (2016). Woodhead Publishing.
8. "An Updated Lifecycle Assessment for Disposable and Reusable Nappies." Environment Agency. Bristol, UK (2008).
9. Berners-Lee, M. *How Bad Are Bananas? The Carbon Footprint of Everything.* Profile Books (2020), London.
10. Retrieved from http://www.madsci.org/posts/archives/1999-03/921040790.Ch.r.html.
11. Shramko, A., Shramko, A., and Shramko, A. "Which Diaper Is More Absorbent, Huggies or Pampers?" *Journal of Emerging Investigators* (2013): https://www.emerginginvestigators.org/articles/which-diaper-is-more-absorbent-huggies-or-pampers.
12. Sen, P., Kantareddy, S.N.R., Bhattacharyya, R., et al. "Low-Cost Diaper Wetness Detection Using Hydrogel-Based RFID Tags." *IEEE Sensors Journal* 20, no.6 (2020): 3293–3302.
13. Retrieved from https://patents.google.com/patent/US10034582B2/en.

CHAPTER 8

1. Yildirim, D., Ozyurek, S.E., Ekiz, A., et al. "Comparison of Active vs. Expectant Management of the Third Stage of Labor in Women with Low Risk of Postpartum Hemorrhage: A Randomized Controlled Trial." *Ginekologia Polska* 87, no.5 (2016): 399–404.
2. Liu, X., Ouyang, J.F., Rossello, F.J., et al. "Reprogramming Roadmap Reveals Route to Human Induced Trophoblast Stem Cells." *Nature* 586 (2020): 101–107.

3. James, J.L., Chamley, L.W., and Clark, A.R. "Feeding Your Baby in Utero: How the Uteroplacental Circulation Impacts Pregnancy." *Physiology* 32 (2017): 234–245.
4. Romo, A., Carceller, R., and Tobajas, J. "Intrauterine Growth Retardation (IUGR): Epidemiology and Etiology." *Pediatric Endocrinology Reviews* 6, supplement 3 (2009): 332–336.
5. Hamilton, W.J., and Boyd, J.D. "Trophoblast in Human Utero-Placental Arteries." *Nature* 212 (1966): 906–908.
6. James, J.L., Saghain, R., Perwick, R., et al. "Trophoblast Plugs: Impact on Utero-Placental Haemodynamics and Spiral Artery Remodelling." *Human Reproduction* 33, no.8 (2018): 1430–1441.
7. Hustin, J., and Schaaps J.P. "Echocardiographic and Anatomic Studies of the Maternotrophoblastic Border During the First Trimester of Pregnancy." *American Journal of Obstetrics and Gynecology* 157 (1987): 162–168.
8. Saghian, R., Bogle, G., James, J.L., et al. "Establishment of Maternal Blood Supply to the Placenta: Insights into Plugging, Unplugging and Trophoblast Behaviour from the Agent-Based Model." *Interface Focus* 9 (2019): 20190019.
9. Allerkamp, H.H., Clark, A.R., Lee, T.C., et al. "Something Old, Something New: Digital Quantification of Uterine Vascular Remodelling and Trophoblast Plugging in Historical Collections Provides New Insight into Adaptation of the Utero-Placental Circulation." *Human Reproduction* 36, no.3 (2021): 571–586.
10. Assali, N.S., Douglass, R.A., Baird, W.W., et al. "Measurement of the Uterine Blood Flow and Uterine Metabolism. IV. Results in Normal Pregnancy." *American Journal of Obstetrics and Gynecology* 66 (1953): 248–253.
11. Plitman Mayo, R., Charnock-Jones, D.S., Burton, G.J., et al. "Three-Dimensional Modeling of the Human Placental Terminal Villi." *Placenta* 43 (2016): 54–60.
12. Jensen, O.E., and Chernyavsky, I.L. "Blood Flow and Transport in the Human Placenta." *Annuals Review of Fluid Mechanics* 51 (2019): 25–47.
13. Inger, G.R. "Scaling Nonequilibrium-Reacting Flows: the Legacy of Gerhard Damköhler." *Journal of Spacecraft and Rockets* 38 (2001): 185.
14. Erlich, A., Pearce, P., Plitman Mayo, R., et al. "Physical and Geometric Determinants of Transport in Fetoplacental Microvascular Networks." *Science Advances* 5 (2019): eaav6326.
15. Carter, A.M. "Animal Models of Human Placentation—A Review." *Placenta* 28 (2007): S41–S47.
16. Dellschaft, N.S., Hutchinson, G., Shah, S., et al. "The Haemodynamics of the Human Placenta in Utero." *PLoS Biology* 18, no.5 (2020): e3000676.
17. Stillbirth Collaborative Research Network. "Causes of Death Among Stillbirths" *Journal of the American Medical Association* 306 (2011):2459–2468.
18. Retrieved from https://www.marchofdimes.org/complications/placental-abruption.aspx.
19. Partridge, E.A., Davey, M.G., Hornick, M.A., et al. "An Extra-uterine System to Physiologically Support the Extreme Premature Lamb." *Nature Communications* 8 (2017): 15112.

20. Costeloe, K.L., Hennessy, E.M., Haider, S., et al. "Short Term Outcomes After Extreme Preterm Birth in England: Comparison of Two Birth Cohorts in 1995 and 2006." *British Medical Journal* 345 (2012): e7976.
21. Retrieved from https://www.bpas.org/get-involved/campaigns/briefings/premature-babies/.
22. Bové, H., Bongaerts, E., Slenders, E., et al. "Ambient Black Carbon Particles Reach the Fetal Side of the Human Placenta." *Nature Communications* 10 (2019): 3866.
23. Ragusa, A., Svelato, A., Santacroce, C., et al. "Plasticenta: First Evidence of Microplastics in Human Placenta." *Environment International* (2021): https://doi.org/10.1016/j.envint.2020.106274.

CHAPTER 9

1. Levington, S. "For John and Jackie Kennedy, the Death of a Son May Have Brought Them Closer." *Washington Post* (2013): https://www.washingtonpost.com/opinions/for-john-and-jackie-kennedy-the-death-of-a-son-may-have-brought-them-closer/2013/10/24/2506051e-369b-11e3-ae46-e4248e75c8ea_story.html.
2. Wrobel, S., and Clements, J.A. "Bubbles, Babies and Biology: The Story of Surfactant." *Federation of American Societies of Experimental Biology* 18 (2004): 1624e. doi:10.1096/fj.04-2077bkt.
3. Singhal, N., Lockyer, J., Fidler, H., et al. "Helping Babies Breath: A Global Neonatal Resuscitation Program Development and Formative Educational Evaluation." *Resuscitation* 83, no.1 (2012): 90–96.
4. Blank, D.A., Gaertner, V.D., Kamlin, C.O.F., et al. "Respiratory Changes in Term Infants Immediately After Birth." *Resuscitation* 130 (2018): 105–110.
5. "Kennedys Mourning Baby Son; Funeral Today Will Be Private." *New York Times*, August 10, 1963, Page 1.
6. Robinson, A. *The Last Man Who Knew Everything.* (2006), PiPress, ISBN 1851684948.
7. Young, T. "III. An Essay on the Cohesion of Fluids." *Philosophical Transactions of the Royal Society* 95 (1805): 65–87.
8. Laplace, P.S. Supplement to the 10th edition of *Mécanique Céleste* (Paris, France: Courcier, 1806).
9. von Neergaard, K. "Neue auffassungen uber einen grundbegriff der atemmechanik.Die retraktionskraft der lunge, abhangig von der oberflachenspannung in denalveolen." *Zeitschrift für die Gesamt Experimentelle Medizine* 66 (1929): 373–394.
10. Pattle, R.E. "Properties, Function and Origin of the Alveolar Lining Layer." *Nature* 175 (1955): 1125–1126.
11. Pattle, R.E. "Properties, Function and Origin of the Alveolar Lining Layer." *Proceedings of the Royal Society B: Biological Sciences* 148 (1958): 217–240.
12. Clements, J.A. "Dependence of Pressure-Volume Characteristics of Lungs on Intrinsic Surface-Active Material." *American Journal of Physiology* 187 (1956): 592.

13. Avery, M.E., and Mead, J. "Surface Properties in Relation to Atelectasis and Hyaline Membrane." *American Journal of Disease in Children* 97 (1959): 517–523.
14. Chakraborty, M., and Kotecha, S. "Pulmonary Surfactant in Newborn Infants and Children." *Breathe* 9, no.6 (2013): 477–488.
15. Nkadi, P.O., Allen Merritt, T., and Pillers, D-A.M. "An Overview of Pulmonary Surfactant in the Neonate: Genetics, Metabolism and the Role of Surfactant in Health and Disease." *Molecular Genetics and Metabolism* 97, no.2 (2009): 95–101.
16. Gregory G.A., Kitterman, J.A., Phibbs, R.H., *et al.* "Treatment of the Idiopathic Respiratory-Distress Syndrome with Continuous Positive Airway Pressure." *New England Journal of Medicine* 284 *(*1971*)*:1333–1340.
17. Fujiwara, T., Chida, S., Watabe, Y., et al. "Artificial Surfactant Therapy in Hyaline-membrane Disease." *The Lancet* 315 (1980): 55–59.
18. Halliday, H.L. "Surfactants: Past, Present and Future." *Journal of Perinatology* 28 (2008): S47–S56.
19. Shaffer, T.H., Wolfson, M.R., and Greenspan, J.S. "Liquid Ventilation: Current Status." *Pediatrics in Review* 20, no 12 (1999): e134–e142.
20. Filoche, M., Tai, C-F., and Grotberg, J.B. "Three-Dimensional Model of Surfactant Replacement Therapy." *Proceedings of National Academies of Science* 112, no.3 (2015): 9287–9292.
21. Kazemi, A., Louis, B., Isabey, D., et al. "Surfactant Delivery in Rat Lungs: Comparing 3D Geometrical Simulation Model with Experimental Instillation." *PLoS Computational Biology* 15, no.10 (2019): e1007408.
22. Copploe, A., Vatani, M., Choi, J-W., et al. "A Three-Dimensional Model of Human Lung Airway Tree to Study Therapeutics Delivery in the Lungs." *Annals of Biomedical Engineering* 47 (2019): 1435–1445.
23. Dabaghi, M., Rochow, N., Saraei, N., et al. "A Pumpless Microfluidic Neonatal Lung Assist Device for Support of Preterm Neonates in Respiratory Distress." *Advanced Science* 21, no.7 (2020): 2001860.
24. Blank, D.A., Rogerson, S.R., Kamlin, C.O.F., et al. "Lung Ultrasound During the Initiation of Breathing in Healthy Term and Late Preterm Infants Immediately After Birth, a Prospective, Observational Study." *Resuscitation* 114 (2017): 59–64.

CHAPTER 10

1. Wolke, D., Bilgin, A., and Samara, M. "Systematic Review and Meta-analysis: Fussing and Crying Durations and Prevalence of Colic in Infants." *Journal of Pediatrics* 185 (2017): 55–61.e4.
2. Lingle, S., and Riede, T. "Deer Mothers Are Sensitive to Infant Distress Vocalizations of Diverse Mammalian Species." *American Naturalist* 184 (2014): 510–522.
3. Marlin, B.J., Mitre, M., D'amour, J.A., et al. "Oxytocin Enables Maternal Behaviour by Balancing Cortical Inhibition." *Nature* 52 (2015): 499–504.
4. Hernandez-Miranda, L.R., Ruffault, P-L., Bouvier, J.C., et al, "Genetic Identification of a Hindbrain Nucleus Essential for Innate Vocalization." *Proceedings of the National Academy of Sciences* 114 (2017): 8095–8100.

5. Bornstein, M.H., Putnick, D.L., Rigo, P., et al. "Neurobiology of Culturally Common Maternal Responses to Infant Cry." *Proceedings of the National Academy of Sciences* 114 (2017): E9465–E9473.
6. De Pisapia, N., Bornstein, M.H., Rigo, P., et al. "Gender Differences in Directional Brain Responses to Infant Hunger Cries." *Neuroreport* 24 (2013): 142–146.
7. Klemuk, S.A., Riede, T., Walsh, E.J., et al. "Adapted to Roar: Functional Morphology of Tiger and Lion Vocal Folds." *PLoS One* 6, no.11 (2011): e27029.
8. May, R.M. "Simple Mathematics Models with Very Complicated Dynamics." *Nature* 261 (1976):459–467.
9. For a review, see: Tecumseh Fitch, W., Neubauer, J., and Herzel. H. "Calls Out of Chaos: The Adaptive Significance of Nonlinear Phenomena in Mammalian Vocal Production." *Animal Behaviour* 63 (2002): 407–418.
10. Mende, W., Herzel, H., and Wermke, K. "Bifurications and Chaos in Newborn Infant Cries." *Physics Letters A* 145 (1990): 418.
11. Robb, M. "Bifucations and Chaos in the Cries of Full-Term and Preterm Infants." *Folia Phoniatr Logop* 55 (2003): 233–240.
12. Fuamenya, N.A., Robb, M., and Wermke, K. "Noisy but Effective: Crying Across the First 3 Months." *Journal of Voice* 29 (2015): 281.
13. Blumstein, D.T., Bryant, G.A., and Kaye, P. "The Sound of Arousal in Music Is Context-Dependent." *Biology Letters* 8 (2012): 744–747.

CHAPTER 11

1. Fredeen, R.C. "Cup Feeding of Newborn Infants." *Pediatrics* 2, no.5 (1948): 544–548.
2. Retrieved from https://www.who.int/news/item/15-01-2011-exclusive-breastfeeding-for-six-months-best-for-babies-everywhere.
3. Victoria, C.G., Bahl, R., Barros, A.J.D., et al. "Breastfeeding in the 21st Century: Epidemiology, Mechanisms, and Lifelong Effect." *The Lancet* 387 (2016): 475–490.
4. Retrieved from https://www.nih.gov/news-events/nih-research-matters/breastfeeding-may-help-prevent-type-2-diabetes-after-gestational-diabetes.
5. Boss, M., Gardner, H., and Hartmann, P. "Normal Human Lactation: Closing the Gap." *F1000Research* 7 (2018): F1000 Faculty-Rev-801.
6. Kent, J.C., Mitoulas, L.R., Cregan, M.D., et al. "Volume and Frequency of Breastfeeding and Fat Content of Breast Milk Throughout the Day." *Pediatrics* 117 (2006): e387–e395.
7. Kent, J.C., Gardner, H., and Geddes, D.T. "Breastmilk Production in the First 4 Weeks After Birth of Term Infants." *Nutrients* 8 (2016): 756.
8. Martin, C.R., Ling, P-R., and Blackburn, G.L. "Review of Infant Feeding: Key Features of Breast Milk and Infant Formula." *Nutrients* 8 (2016): 279. doi:10.3390/nu8050279.
9. Mortazavi, S.N., Geddes, D.T., and Hassanipour, F. "Lactation in the Human Breast from a Fluid Dynamics Point of View." *Journal of Biomechanical Engineering* 139 (2017): 011009.

10. Wiciński, M., Sawicka, E., Gębalski, J., et al. "Human Milk Oligosaccharides: Health Benefits, Potential Applications in Infant Formulas, and Pharmacology." *Nutrients* 12 (2020): 266. doi:10.3390/nu12010266.
11. Newburg, D.S., Ruiz-Palacios, G.M., and Morrow, A.L. "Human Milk Glycans Protect Infants Against Enteric Pathogens." *Annual Review of Nutrition* 25 (2005): 37–58.
12. Ribo, S., Sanchez-Infantes, D., Martinex-Guino, L., et al. "Increasing Breast Milk Betaine Modulates *Akkermansia* Abundance in Mammalian Neonates and Improves Long-Term Metabolic Health." *Science Translational Medicine* 13 (2021): eabb0322.
13. Wagner, E.A., Chantry, C.J., Dewey, K.G., et al. "Breastfeeding Concerns at 3 and 7 Days Postpartum and Feeding Status at 2 Months." *Pediatrics* 132, no.4 (2013): e865–e875.
14. Garner, C.D., Ratcliff, S.L., Thornburg, L.L., et al. "Discontinuity of Breastfeeding Care: There's No Captain of the Ship." *Breastfeeding Medicine* 11, no.1 (2016): 32–39.
15. "Breastfeeding and Breast Milk—from Biochemistry to Impact." Georg Thieme Verlag (2018) Stuttgart.
16. Cooper, A.P. "On the Anatomy of the Breast." Longman (1840): https://jdc.jefferson.edu/cooper/.
17. Basch, K. "Beiträge zur Kenntniss des menschlichen Milchapparats." *Archiv für Gynäkologie* 44 (1893):15–54.
18. Kron, R.E., and Litt, M. "Fluid Mechanics of Nutritive Sucking Behaviour: The Suckling of Infant's Oral Apparatus Analysed as a Hydraulic Pump." *Medical & Biological Engineering* 9 (1971): 45–60.
19. Von Pfaundler, M. "Über Saugen and Verdauen." *Ver Ges Kinderheil* 16 (1899): 38–53.
20. Hytten, F.E. "Observation on the Vitality of the Newborn." *Archives of Disease in Childhood* 26 (1951): 477–486.
21. Ardran, G.M., Kemp, F.H., and Lind, J. "A Cineradiographic Study of Breast Feeding." *British Journal of Radiology* 31 (1958): 156–162.
22. Woolridge, M.W. "The 'Anatomy' of Infant Sucking." *Midwifery* 2 (1986): 164–171.
23. Eishima, K. "The Analysis of Sucking Behaviour in Newborn Infants." *Early Human Development* 27 (1991): 163–173.
24. Smith, W.L., Erenber, A., Nowak, A., et al. "Physiology of Sucking in the Normal Term Infant Using Real-Time US." *Radiology* 156 (1985): 379–381.
25. Smith, W.L., Erenberg, A., and Nowak, A. "Imaging Evaluation of the Human Nipple During Breast-Feeding." *American Journal of Diseases of Children* 142 (1988): 76–78.
26. Geddes, D.T., Kent, J.C., Mitoulas, L.R., et al. "Tongue Movement and Intra-oral Vacuum in Breastfeeding Infants." *Early Human Development* 84 (2008): 471–477.
27. Sakalidis, V.S., Williams, T.M., Garbin, C.P., et al. "Ultrasound Imaging of Infant Sucking Dynamics During the Establishment of Lactation. "*Journal of Human Lactation* 29, no.2 (2013): 205–213.

28. O'Shea, J.E., Foster, J.P., O'Donnell, C.P.F., et al. "Frenotomy for Tongue-Tie in Newborn Infants." *Cochrane Database of Systematic Reviews* Issue 3 (2017): CD011065.
29. Elad, D., Kozlovsky, P., Blum, O., et al. "Biomechanics of Milk Extraction During Breast-Feeding." *Proceedings of the National Academy of Sciences* 111 (2014): 5230–5235.
30. Bu'Lock, F., Woolridge, M.W., and Baun, J.D. "Development of Co-ordination of Sucking, Swallowing and Breathing: Ultrasound Study of Term and Preterm Infants." *Developmental Medicine and Child Neurology* 32 (1990): 669–678.
31. Monaci, G., and Woolridge, M. "Ultrasound Video Analysis for Understanding Infant Breastfeeding." *Proceedings of the 18th IEEE International Conference on Image Processing* (2011): 1765–1768. doi: 10.1109/ICIP.2011.6115802.
32. Genna, C.W., Saperstein, Y., Siegal, S.A., et al. "Quantitative Imaging of Tongue Kinematics During Infant Feeding and Adult Swallowing Reveals Highly Conserved Patterns." *Physiological Reports* 9 no.3 (2021): ee14685.

THIRD INTERLUDE

1. Obladen M. "Guttus, Tiralatte and Téterelle: A History of Breast Pumps." *Journal of Perinatal Medicine* 40, no.6 (2012): 669–675.
2. Retrieved from https://makethebreastpumpnotsuck.com/.
3. Kent, J.C., Mitoulas, L.R., Cregan, M.D., et al. "Importance of Vacuum for Breastmilk Expression." *Breastfeeding Medicine* 3, no.1 (2008): 11–19.
4. Dunne, J.B., Rebay-Salisbury, K., Salisbury, K.R., et al. "Milk of Ruminants in Ceramic Baby Bottles from Prehistoric Child Graves." *Nature* 574 (2019): 246–248.
5. Stevens, E.E., Patrick, T. E., Pickler, R. "A History of Infant Feeding." *Journal of Perinatal Education* 18, no.2 (2009): 32–39.
6. Radbill, S. X "Infant Feeding Through the Ages." *Clinical Pediatrics* 20, no.10 (1981): 613–621.
7. Wickes, I.G. "A History of Infant Feeding. Part I. Primitive Peoples: Ancient Works: Renaissance Writers." *Archives of Disease in Childhood* 28 (1953): 232–240.
8. Wickes, I.G. "A History of Infant Feeding. Part III: Eighteenth and Nineteenth Century Writers." *Archives of Disease in Childhood* 28 (1953): 332340.
9. Weinberg, F. "Infant Feeding Through the Ages." *Canadian Family Physician* 39 (1993): 2016–2020.
10. Wickes, I.G. "A History of Infant Feeding. Part IV: Nineteenth Century Continued." *Archives of Disease in Childhood* 28 (1953): 416–422.

CHAPTER 12

1. Iglowstein, I., Jenni, O.G., Molinari, L., et al. "Sleep Duration from Infancy to Adolescence: Reference Values and Generational Trends." *Pediatrics* 111, no.2 (2003): 302–307.

2. Rivkees, S.A., Mayes, L., Jacobs, H., et al. "Rest-Activity Patterns of Premature Infants Regulated by Cycled Light." *Pediatrics* 113, no.4 (2004): 833–839.
3. Wielek, T., Del Giudice, R., Lang, A., et al. "On the Development of Sleep States in the First Weeks of Life." *PLoS ONE* 14, no.10 (2019): e0224521.
4. Grigg-Damberger, M.M. "The Visual Scoring of Sleep in Infants 0 to 2 Months of Age." *Journal of Clinical Sleep Medicine* 12, no.3 (2016): 429–445.
5. Cao, J., Herman, A.B., West, G.B., et al. "Unravelling Why We Sleep: Quantitative Analysis Reveals Abrupt Transition from Neural Reorganisation to Repair in Early Development." *Science Advances* 6 (2020): eaba0398.
6. Lo, C-C., Nunes Amaral, L.A., Havlin, S., et al. "Dynamics of Sleep-Wake Transitions During Sleep." *Europhysics Letters* 57, no 5 (2002): 626–631.
7. Lo, C-C., Chou, T., Penzel, T., et al. "Common Scale-Invariant Patterns of Sleep-Wake Transitions Across Mammalian Species." *Proceedings of the National Academies of Sciences* 101 (2004) 17545–17548.
8. Retrieved from https://safetosleep.nichd.nih.gov/safesleepbasics/risk/factors.
9. Retrieved from https://www.cdc.gov/sids/about/index.htm.
10. Task Force on Sudden Infant Death Syndrome. "The Changing Concept of Sudden Infant Death Syndrome: Diagnostic Coding Shifts, Controversies Regarding the Sleeping Environment, and New Variable to Consider in Reducing Risk." *Pediatrics* 116 (2005):1245–1255.
11. Saper, C. B., Scammell, T. E., and Lu, J. "Hypothalamic Regulation of Sleep and Circadian Rhythms." *Nature* 437 (2005): 1257–1263.
12. Dvir, H., Elbaz, I., Havlin, S., et al. "Neuronal Noise as an Origin of Sleep Arousals and Its Role in Sudden Infant Death Syndrome." *Science Advances* 4 (2018): eaar6277.
13. Steinmetz, P., Manwani, A., Kock, C., et al. "Subthreshold Voltage Noise Due to Channel Fluctuations in Active Neuronal Membranes." *Journal of Computational Neuroscience* 9 (2000): 133–148.
14. Wailoo, M.P., Petersen, S.A., Whittaker, H., et al. "Sleeping Body Temperatures in 3–4 Month Old Infants." *Archives of Disease in Childhood* 64 (1989): 596–599.
15. Whitehead, K., Pressler, R., and Fabrizi, L. "Characteristics and Clinical Significance of Delta Brushes in EEG of Premature Infants." *Clinical Neurophysiology Practice* 2 (2017): 12–18.
16. Johnson, J.B. "The Schottky Effect in Low-Frequency Circuits." *Physical Review* 26 (1925): 71–85.
17. Schottky, W. "Small-Shot Effect and Flicker Effect." *Physical Review* 28 (1926): 74–103.
18. Voss, R.F., and Clarke, J. "'1/f Noise' in Music and Speech." *Nature* 258 (1975): 317–318.
19. Xiao, R., Shida-Tokeshi, J., Vanderbilt, D.L., et al. "Electroencephalography Power and Coherence Changes with Age and Motor Skill Development Across the First Half Year of Life." *PLOS One* 13 (2018): e0190276.
20. Schaworonkow, N., and Voytek, B. "Longitudinal Changes in Aperiodic and Periodic Activity in Electrophysiological Recordings in the First Seven Months of Life." *Developmental Cognitive Neuroscience* 47 (2021): 100895.

21. Voytek, B., Kramer, M.A., Case, J., et al. "Age-Related Changes in 1/f Neural Electrophysiological Noise." *Journal of Neuroscience* 35 (2015): 13257–13265.
22. Demene, C., Baranger. J., Bernal, M., et al. "Functional Ultrasound Imaging of Brain Activity in Human Newborns." *Science Translational Medicine* 9 (2017): eaah6756.
23. Hill, R.M., Boto, E., Holmes, N., et al. "A Tool for Functional Brain Imaging with Lifespan Compliance." *Nature Communications* 10 (2019): 4785.

CHAPTER 13

1. Adolph, K.E., Vereijken, B., and Denny, M.A. "Learning to Crawl." *Child Development* 69, no.5 (1998) 1299–1312.
2. Righetti, R., Nylén, A., Rosander, K., et al. "Kinematic and Gait Similarities Between Crawling Human Infants and Other Quadruped Mammals." *Frontiers in Neurology* 6 (2015) 1–17: https://doi.org/10.3389/fneur.2015.00017.
3. Zelazo, P.R. "The Development of Walking: New Findings and Old Assumptions." *Journal of Motor Behaviour* 15, no.2 (1983): 99–137.
4. Dominici, N., Ivanenko, Y.P., Cappellini, G., et al. "Locomotor Primitives in Newborn Babies and Their Development." *Science* 334 (2011): 997–999.
5. Sylos-Labini, F., La Scaleia, V., Cappellini, G., et al. "Distinct Locomotor Precursors in Newborn Babies." *Proceedings of the National Academies of Science* 117, no.17 (2020): 9604–9612.
6. Del Vecchio, A., Sylos-Labini, F., Mondi, V., et al. "Spinal Motoneurons of the Human Newborn Are Highly Synchronized During Leg Movements." *Science Advances* 6 (2020): eabc3916.
7. Barbu-Roth, M., Anderson, D.I., Desprès, A., et al. "Air Stepping in Response to Optic Flows That Move Toward and Away from the Neonate." *Developmental Psychobiology* 56, no.5 (2014): 1142–1149.
8. Forma, V., Anderson, D.I., Goffinet, F., et al. "Effect of Optic Flows on Newborn Crawling." *Developmental Psychobiology* 60, no.5 (2018): 497–510.
9. Forma, V., Anderson, D.I., Provasi, J., et al. "What Does Prone Skateboarding in the Newborn Tell Us About the Ontogeny of Human Locomotion?" *Child Development* 90, no.4 (2019): 1286–1302.
10. Hym, C., Forma, V., Anderson, D.I., et al. "Newborn Crawling and Rooting in Response to Maternal Breast Odor." *Developmental Science* (2020): https://doi.org/10.1111/desc.13061.
11. Alexander, R.M. "Optimization and Gaits in the Locomotion of Vertebrates." *Physiological Reviews* 69, no.4 (1989): 1199–1227.
12. Cavagna, G.A., Thys, H., and Zamboni, A. "The Sources of External Work in Level Walking and Running." *Journal of Physiology* 262 (1976): 639–657.
13. Cavagna, G.A., Franzetti, P., and Fuchimoto, T. "The Mechanics of Walking in Children." *Journal of Physiology* 343 (1983): 323–339.
14. Ivanenko, Y.P., Dominici, N., Cappellini, G., et al. "Developmental of Pendulum Mechanism and Kinematic Coordination from the First Unsupported Steps in Toddlers." *Journal of Experimental Biology* 2017 (2004): 3797–3810.

15. Ivanenko, Y.P., Dominici, N., and Lacquaniti, F., et al. "Development of Independent Walking in Toddlers." *Exercise and Sport Sciences Reviews* 35, no.2 (2007): 67–73.
16. Muir, G.D., Gosline, J.M., and Steeves, J.D. "Ontogeny of Bipedal Locomotion: Walking and Running in the Chick." *Journal of Physiology* 493 (1996): 589–601.
17. Ivanenko, Y.P., Dominici, N., Cappellini, G., et al. "Kinematics in Newly Walking Toddlers Does Not Depend Upon Postural Stability." *Journal of Neurophysiology* 94 (2004): 754–763.
18. Dominici, N., Ivanenko, Y.P., Cappellini, G., et al. "Kinematic Strategies in Newly Walking Toddlers Stepping Over Different Support Surfaces." *Journal of Neurophysiology* 103 (2010): 1673–1684.
19. Marencakova, J., Price, C., Maly, T., et al. "How Do Novice and Improver Walkers Move in Their Home Environments? An Open-Sourced Infant's Gait Video Analysis." *PLOS One* 14, no.6 (2019): e0218665.
20. Walle, E., and Campos, J. "Infant Language Development Is Related to the Acquisition of Walking." *Developmental Psychology* 50, no.2 (2014):336–348.

FOURTH INTERLUDE

1. Pendrill, A-M., and Williams, G. "Swings and Slides." *Physics Education* 40, no.6 (2005): 527–533.
2. Case, W.B., and Swanson, M.A. "The Pumping of a Swing from the Seated Position." *American Journal of Physics* 58 (1990): 463–467.
3. Roura, P., and González, J.A. "Towards a More Realistic Description of Swing Pumping Due to the Exchange of Angular Momentum." *European Journal of Physics* 31 (2010): 1195–1207.
4. Case, W.B. "The Pumping of a Swing from a Standing Position." *American Journal of Physics* 63 (1996): 215–220.
5. Wirkus, S., Rand, R., and Ruina, A. "How to Pump a Swing." *College Mathematics Journal* 29, no.4 (1998): 266–275.
6. Thompson, M., Barron, P., Chandler, C., et al. "Playground Fun Demonstrates Rotational Mechanics Concepts." *Physics Education* 45 (2010): 459–461.
7. Pendrill, A-M., and Eager, D. "Free Fall and Harmonic Oscillations: Analyzing Trampoline Jumps." *Physics Education* 50 (2015): 64–70.
8. Güémez, J., Fiolhais, C., and Fiolhais, M. "Toys in Physics Lectures and Demonstrations—a Brief Review." *Physics Education* 44 (2009): 53–64.

CHAPTER 14

1. Ota, M., Davies-Jenkins, N., and Skarabela, B. "Why Choo-Choo Is Better Than Train: The Role of Register-Specific Words in Early Vocabulary Growth." *Cognitive Science* 42, no 6 (2019): 1974–1999.
2. Masapollo, M., Polka, L., and Menard, L. "When Infants Talk, Infants Listen: Pre-babbling Infants Prefer Listening to Speech with Infant Vocal Properties." *Developmental Science* 19, no. 2 (2015): 318–328.

3. Mampe, B., Friederici, A.D., Christophe, A., et al. "Newborns' Cry Melody Is Shaped by Their Native Language." *Current Biology* 19 (2009): P1994–1997.
4. Bergelson, E., and Aslin, R. "Nature and Origins of the Lexicon in 6-Mo-Olds." *Proceedings of the National Academies of Science* 114 (2017): 12916–12921.
5. Saffron, J.R. "Statistical Language Learning in Infancy." *Child Development Perspectives* 14 (2020): 49–54.
6. Wermke, K., Robb, M.P., and Schluter, P.J. "Melody Complexity of Infants' Cry and Non-cry Vocalisations Increases Across the First Six Months." *Scientific Reports* 11 (2021): 4137.
7. Skarabela, B., Ota, M., O'Connor, R., et al. "'Clap Your Hands' or 'Take Your Hands?' One-Year-Olds Distinguish Between Frequent and Infrequent Multiword Phrases." *Cognition* 211 (2021): 104612.
8. Schneider, R., Yorovsky, D., and Frank, M. "Large-Scale Investigations of Variability in Children's First Words." Proceedings of the 37th Annual Conference of the Cognitive Science Society (2015): 2210–2115: https://cogsci.mindmodeling.org/2015/papers/0364/index.html.
9. Retrieved from https://wordbank.stanford.edu/.
10. Retrieved from https://langcog.github.io/wordbank-book/items-consistency.html#the-first-10-words.
11. Fenson, L., Dale, P.S., and Reznick, J.S. "Variability in Early Communicative Development." Monographs of the Society for research in Child Development." 59, no.5 (1994): 1–173.
12. Roy, B.C., Frank, M.C., DeCamp, P., et al. "Predicting the Birth of a Spoken Word." *Proceedings of the National Academies of Sciences* 112, no.4 (2015): 12663–12668.
13. Acquired from: https://www.ethnologue.com/guides/how-many-languages.
14. Zipf, G.K. *Human Behavior and the Principle of Least Effort: An Introduction to Human Ecology.* Hafner reprint (1972), New York.
15. Yang, C., Crain, S., Berwick, R.C., et al. "The Growth of Language: Universal Grammar, Experience, and Principles of Computation." *Neuroscience and Biobehavioural Reviews* 81 (2017): 103–119.
16. Chomsky, N. "Three Models for the Description of Language," in *IRE Transactions on Information Theory*, vol. 2, no. 3 (1956): 113–124, doi: 10.1109/TIT.1956.1056813.
17. Hamburger, H., and Crain, S. "Acquisition of Cognitive Compiling." *Cognition* 17 (1984): 85–136.
18. De Giuli, E. "Random Language Model." *Physical Review Letters* 122 (2019): 128301.
19. Retrieved from https://childes.talkbank.org/.
20. Corominas-Murtra, B., Valverde, S., and Solé, R. "The Ontogeny of Scale-Free Syntax Networks: Phase Transitions in Early Language Acquisition." *Advances in Complex Systems* 12, no.3 (2009): 371–392.

AFTERWORD

1. Retrieved from: https://www.theguardian.com/education/2018/aug/10/female-scientists-urge-research-grants-reform-tackle-gender-bias.

INDEX

Page numbers in *italics* indicate illustrations

C

ABOUT THE AUTHOR

Michael Banks was born in Oldham, Lancashire, UK. After an undergraduate degree in physics from Loughborough University, UK, Michael did a PhD in condensed-matter physics at the Max Planck Institute for Solid State Physics in Stuttgart, Germany, studying magnetism.

For over a decade, Michael has been news editor of the international monthly magazine *Physics World*, where he covers the latest developments in physics. In addition to *Physics World*, Michael has written for *Nature*, *BBC Science Focus*, and *Science Uncovered*, as well as appeared on BBC Radio 4. Michael is based in Bristol, UK, where he lives with his wife and two boys. He tweets at @Mike_Banks.